Synthesis Lectures on Computer Science

The series publishes short books on general computer science topics that will appeal to advanced students, researchers, and practitioners in a variety of areas within computer science.

Deepshikha Bhati · Fnu Neha ·
Angela Guercio · Md Amiruzzaman ·
Aloysius Bathi Kasturiarachi

A Beginner's Guide to Generative AI

An Introductory Path to Diffusion Models, ChatGPT, and LLMs

Deepshikha Bhati
Department of Computer Science
Kent State University
North Canton, OH, USA

Fnu Neha
Department of Computer Science
Kent State University
Kent, OH, USA

Angela Guercio
Department of Computer Science
Kent State University
North Canton, OH, USA

Md Amiruzzaman
Department of Computer Science
West Chester University
West Chester, PA, USA

Aloysius Bathi Kasturiarachi
Department of Mathematical Sciences
Kent State University
North Canton, OH, USA

ISSN 1932-1228   ISSN 1932-1686 (electronic)
Synthesis Lectures on Computer Science
ISBN 978-3-031-84723-3   ISBN 978-3-031-84724-0 (eBook)
https://doi.org/10.1007/978-3-031-84724-0

© The Editor(s) (if applicable) and The Author(s), under exclusive license to Springer Nature Switzerland AG 2026

This work is subject to copyright. All rights are solely and exclusively licensed by the Publisher, whether the whole or part of the material is concerned, specifically the rights of translation, reprinting, reuse of illustrations, recitation, broadcasting, reproduction on microfilms or in any other physical way, and transmission or information storage and retrieval, electronic adaptation, computer software, or by similar or dissimilar methodology now known or hereafter developed.

The use of general descriptive names, registered names, trademarks, service marks, etc. in this publication does not imply, even in the absence of a specific statement, that such names are exempt from the relevant protective laws and regulations and therefore free for general use.

The publisher, the authors and the editors are safe to assume that the advice and information in this book are believed to be true and accurate at the date of publication. Neither the publisher nor the authors or the editors give a warranty, expressed or implied, with respect to the material contained herein or for any errors or omissions that may have been made. The publisher remains neutral with regard to jurisdictional claims in published maps and institutional affiliations.

This Springer imprint is published by the registered company Springer Nature Switzerland AG
The registered company address is: Gewerbestrasse 11, 6330 Cham, Switzerland

If disposing of this product, please recycle the paper.

Preface

In this fast-growing environment of creating artificial intelligence (AI)-based applications, generative AI has emerged as one of the groundbreaking technologies of the century. Self-driving cars, healthcare, and other industries are realizing the potential of generative AI through its application possibilities, presenting a disruptive approach that transforms the creations of text, images, and even conversations between people. This book, *A Beginner's Guide to Generative AI: An Introductory Path to Diffusion Models, ChatGPT, and LLMs*, aims to assist you in exploring this fascinating domain while equipping you with the necessary background information and recommendations.

To understand what is happening in this fresh and innovative area for our future development, we must embark on a long and profound journey, beginning by constructing a clear and strong pictures of the fundamental approaches and concepts that define generative AI. Here, you will explore the architectures of some of the most commonly used models, such as Transformers, ChatGPT, and Google Bard, and learn how these technologies have achieved such impressive results. Additionally, you will discover how generative AI can be utilized in writing, medicine, business, and law through concrete examples.

This book is intended for novice readers, and the basics of working with generative AI do not require any prior training. From this perspective, we will proceed from the most basic concepts to more advanced topics in a sequential and accessible manner. Whether you are a student, a working professional, or simply someone curious about what AI can offers you will find this guide highly valuable.

We adopt a broad outlook, presenting a wide array of techniques and models across the spectrum. Rather than focusing on a single method, you will gain insights into generative modeling techniques such as diffusion models, variational autoencoders, and transformers. Understanding all approaches is essential because the field of generative AI is continuously evolving.

As you progress through this book, you will realize that generative AI is not just a series of algorithms; it is an opportunity to create, solve, and innovate. While our primary goal is to enhance your knowledge and awareness about AI, we also hope to inspire your imagination about what AI could become in the future.

Greetings, everyone. You are now in the world of generative AI. Adventure awaits!

North Canton, USA	Deepshikha Bhati
Kent, USA	Fnu Neha
North Canton, USA	Angela Guercio
West Chester, USA	Md Amiruzzaman
North Canton, USA	Aloysius Bathi Kasturiarachi

Competing Interests The authors have no competing interests to declare that are relevant to the content of this manuscript.

Contents

1 Introduction to Generative AI 1
   1.1 Define Generative AI 1
   1.2 The Essence of Generative AI 2
   1.3 Key Components of AI 3
   1.4 Exploring Generative AI Domains 5
      1.4.1 Text Generation 5
      1.4.2 Image Generation 6
      1.4.3 Audio Generation 6
      1.4.4 Video Generation 7
   1.5 Leading Innovators and Their Contributions 8
   1.6 How Does Generative AI Work? 8
   1.7 Generative AI Interfaces 9
   1.8 Recent Transformative Developments in Generative AI 10
   1.9 Current Capabilities and Applications of Generative AI 10
   1.10 Impact on Various Industries 10
   1.11 Comprehensive Overview of Generative AI Applications 12
   1.12 Pros and Cons of Generative AI 13
      1.12.1 Pros 13
      1.12.2 Cons 13
   1.13 Addressing Ethical Considerations 14
   1.14 Summary 14
   1.15 Multiple-choice Questions 14
   1.16 Answers 18
   References 18

2 Evolution of Neural Networks to Large Language Models 21
   2.1 Natural Language Processing 22
   2.2 Probabilistic Models 22
   2.3 N-Gram Models 23

2.4	Hidden Markov Models (HMMs)		24
2.5	Neural Network-Based Language Models		24
2.6	Recurrent Neural Networks (RNNs)		25
	2.6.1	Understanding RNNs and Sequential Data	25
	2.6.2	RNN Architecture	25
	2.6.3	Handling Variable-Length Sequences	26
	2.6.4	Temporal Dependencies and "Memory"	26
	2.6.5	Challenges with RNNs	26
	2.6.6	Limitations of RNNs	27
2.7	Long Short-Term Memory (LSTM) Networks		28
	2.7.1	Key Components of LSTM Networks	28
	2.7.2	LSTM Equations	28
	2.7.3	Applications of LSTM Networks	29
	2.7.4	Advantages of LSTM Networks	30
	2.7.5	Limitations	30
2.8	Gated Recurrent Unit (GRU) Networks		30
	2.8.1	Key Components of GRU Networks	30
	2.8.2	GRU Equations	31
	2.8.3	How GRUs Work?	31
	2.8.4	Advantages of GRU Networks	32
	2.8.5	Applications of GRU Networks	32
	2.8.6	GRU Versus LSTM	33
	2.8.7	Limitations of GRU Networks	33
2.9	Encoder–Decoder Networks		33
	2.9.1	Components of the Encoder–Decoder Architecture	34
	2.9.2	How the Encoder–Decoder Architecture Works?	34
	2.9.3	Key Challenges and Improvements	35
	2.9.4	Applications of the Encoder–Decoder Architecture	36
	2.9.5	Advantages of the Encoder–Decoder Architecture	36
	2.9.6	Limitations	36
	2.9.7	Enhancements and Variations	37
2.10	Attention Mechanism		37
	2.10.1	Traditional Encoder–Decoder Limitation	37
	2.10.2	Introduction of the Attention Mechanism	38
	2.10.3	How the Attention Mechanism Works?	38
	2.10.4	Visualization of Attention	39
	2.10.5	Applications and Impact	39
	2.10.6	Variations and Extensions	40
	2.10.7	Advantages of the Attention Mechanism	41
	2.10.8	Limitations	41

2.11	Transformer Architecture	41
	2.11.1	Components of the Transformer Architecture	41
2.12	Self-Attention Mechanism	42
	2.12.1	Advantages of the Transformer Architecture	43
	2.12.2	Transformer Architecture in Practice	44
2.13	Large Language Models (LLMs)	45
	2.13.1	Transformer Architecture Overview	45
	2.13.2	General Architecture Properties	45
	2.13.3	Examples of Recent LLMs	46
2.14	Summary	47
2.15	Multiple-choice Questions	48
2.16	Answers	50
References	51

3 LLMs and Transformers ... 53

3.1	Unlock the Potential of Language Models	54
3.2	Why Transformers? The Evolution of Neural Architectures	56
	3.2.1	Challenges with RNNs and LSTMs	56
3.3	Inside the Transformer: A Structural Overview	57
	3.3.1	Components of the Transformer Model	57
	3.3.2	Components of the Encoder–Decoder System	58
	3.3.3	Detailed Process	59
	3.3.4	Input Sequence Encoding	60
	3.3.5	Output Sequence Generation	61
3.4	Decoding Attention: The Core of Transformers	62
	3.4.1	Self-Attention Mechanism Explained	62
	3.4.2	Example in Action	64
	3.4.3	Significance of Self-Attention	65
	3.4.4	Code Explanation	67
3.5	Transformers in Perspective: Strengths and Shortcomings	67
	3.5.1	Strengths of the Transformer Model	67
	3.5.2	Shortcomings of the Transformer Model	69
3.6	Related Studies	70
3.7	Summary	71
3.8	Multiple-choice Questions	71
3.9	Answers	74
References	75

4 The ChatGPT Architecture: An In-Depth Exploration of OpenAIs ... 77

4.1	Exploring Conversational Language Models	78
4.2	The Evolution of GPT Models	80
	4.2.1	GPT-1: The Foundation	80

		4.2.2	GPT-2: Scaling Up	81
		4.2.3	GPT-3: The Leap Forward	82
		4.2.4	GPT-4 and Beyond: Pushing Boundaries	83
	4.3	Architecture of ChatGPT		84
		4.3.1	Transformer Architecture: The Backbone	84
		4.3.2	Architecture of ChatGPT: Decoder-Only Structure	85
		4.3.3	Reinforcement Learning from Human Feedback (RLHF)	90
	4.4	Pre-training and Fine-Tuning in ChatGPT		92
		4.4.1	Pre-training	92
		4.4.2	Fine-Tuning: Increase/Decrease Repetition Depending on the Tasks	93
		4.4.3	Continuous Learning and Iterative Improvement	93
	4.5	Contextual Embeddings in ChatGPT		95
		4.5.1	Role of Contextual Embeddings	95
		4.5.2	Generating Contextual Embeddings	96
		4.5.3	Example: Generating a Response for the Input "What Are the Benefits of Regular Exercise?"	99
	4.6	Handling Biases and Ethical Considerations		101
		4.6.1	Addressing Biases in Language Models	101
		4.6.2	Awareness of the Consequences of Biases	101
		4.6.3	Strategies Adopted by OpenAI to Eliminate Biases	101
		4.6.4	Challenges and Trade-Offs	102
		4.6.5	Motivation for Growth	103
	4.7	Strengths and Limitations of ChatGPT		103
		4.7.1	Strengths of ChatGPT	103
		4.7.2	Limitations of ChatGPT	104
	4.8	Related Studies		105
	4.9	Summary		107
	4.10	Multiple-choice Questions		107
	4.11	Answers		110
	References			110
5	**Google Bard and Beyond**			113
	5.1	The Transformer Architecture		114
	5.2	Google Bard's Text and Code Fusion		115
	5.3	Strengths and Weaknesses of Google Bard		116
		5.3.1	Strengths	116
		5.3.2	Weaknesses	116
	5.4	Difference Between ChatGPT and Google Bard		117
	5.5	Claude 2: A Plan For A Qualitative Connection Between Human and Computers		118
		5.5.1	Key Features of Claude 2	118

		5.5.2 Comparing Claude 2 to Other AI Chatbots	119
		5.5.3 The Human-Centered Design Philosophy of Claude	120
		5.5.4 An Analysis of AI Conversational Competencies of Claude	120
	5.6	What is Constitutional AI?	121
	5.7	Claude 2 Versus GPT-3.5	123
	5.8	Other Large Language Models	124
		5.8.1 Falcon AI	124
		5.8.2 LLaMa 2	125
		5.8.3 Dolly 2	127
	5.9	Summary	130
	5.10	Multiple-choice Questions	131
	5.11	Answers	133
	References		134
6	**Diffusion Model and Generative AI for Images**		135
	6.1	Understand the Fundamentals of Variational Autoencoders (VAEs)	137
		6.1.1 Core Concepts of VAEs	137
		6.1.2 Architecture of VAEs	138
		6.1.3 Generative Tasks with VAEs	138
	6.2	Gain Insight Into Generative Adversarial Networks (GANs)	139
		6.2.1 Generative Adversarial Networks (GANs)	139
		6.2.2 Principles Behind GANs	139
		6.2.3 Structure of GANs	140
		6.2.4 Training Process of GANs	140
		6.2.5 Applications in Generating High-Quality Images	141
	6.3	Explore Diffusion Models and Their Types	144
		6.3.1 Diffusion Models	144
		6.3.2 Applications of Diffusion Models	149
		6.3.3 Architecture of Diffusion Models	149
	6.4	Types of Diffusion Model	150
		6.4.1 Denoising Diffusion Probabilistic Models (DDPMs)	150
		6.4.2 Score-Based Diffusion Models (SBMs)	152
	6.5	The Logistics of Understanding DALL-E 2	154
	6.6	Learn About Stable Diffusion and the Latent Diffusion Model (LDM)	156
		6.6.1 Latent Diffusion Model (LDM)	156
		6.6.2 Benefits and Significance	158
	6.7	Find Out What Awaits You During Midjourney: Learn the Technology Behind Midjourney	158
		6.7.1 Generative Adversarial Networks (GANs)	159
		6.7.2 Text-to-Image Synthesis with GANs	159
		6.7.3 Conditional GANs (cGANs)	159

	6.7.4	Training Process	160
	6.7.5	Attention Mechanisms	160
	6.7.6	Data Augmentation and Preprocessing	161
	6.7.7	Benefits and Applications	161
	6.7.8	Example	161
6.8	Why DALL-E 2, Stable Diffusion, and Midjourney Are Not The Same Thing?		163
	6.8.1	DALL·E 2: An Improved Text-to-Image Translation Using CLIP and Diffusion Models	163
6.9	Stable Diffusion: Latent Diffusion Models (LDMs) Toward Making the Business More Efficient		164
6.10	Midjourney: Here, Creative Synthesis with Conditional GANs (cGANs)		165
6.11	Understand the Role of Data Augmentation and Preprocessing		167
	6.11.1	Data Augmentation: Enhancing Model Generalization	167
	6.11.2	Preprocessing: Cleaning Data for the Best Model Training	168
	6.11.3	Synergy Between Augmentation and Preprocessing	169
6.12	Explore the Use of Attention Mechanisms in Image Generation		169
	6.12.1	Understanding Attention Mechanisms	170
	6.12.2	Enhancing Image Generation with Attention	170
	6.12.3	Attention Mechanisms in Diffusion Models	171
	6.12.4	Benefits of Attention Mechanisms in Text-to-Image Models	171
6.13	Learn the Key Evaluation Metrics and Optimization Methods		171
	6.13.1	Role of Loss Functions in Generative Models	172
	6.13.2	GAN-Specific Loss Functions	173
	6.13.3	Optimization Techniques	173
	6.13.4	Techniques to Stabilize GAN Training	174
6.14	Identify the Benefits and Applications of Generative Models		175
	6.14.1	Advantages of Generative Models	175
	6.14.2	Applications of Generative Models	176
6.15	Related Work		177
6.16	Summary		178
6.17	Multiple-choice Questions		179
6.18	Answers		182
References			183

7	Setting Up the Environment and Implementing LLMs	185
	7.1 Environment Setup	185
	7.2 Loading a Pre-Trained LLM	186
	7.3 Generation of Text Using GPT-2	187

	7.4	Fine-Tuning the Model	188
	7.5	Customizing Text Generation	188
	7.6	Summary	190
	7.7	Multiple-choice Questions	190
	7.8	Answers	192
	References		193
8	**ChatGPT Use Cases**		195
	8.1	Business and Customer Service with Chat GPT	195
		8.1.1 Changing the Face of Customer Support	195
		8.1.2 Enhancing Sales and Providing Product	196
		8.1.3 Emphasizing the Continuity of Feedback Analysis	196
		8.1.4 Recommending Personalized Products	197
		8.1.5 Real-Time Order Tracking and Status Updates	197
		8.1.6 Reducing Returns and Refunds	198
	8.2	Content Creation and Marketing with ChatGPT	198
		8.2.1 Blog Post and Article Generation	198
		8.2.2 Social Media Content Creation	199
		8.2.3 SEO-Friendly Content Creation	200
		8.2.4 Communicative Case	200
		8.2.5 Product Descriptions	201
		8.2.6 Consistent Brand Messaging and Tone	201
	8.3	Computer Programming and Technologies Assistance with ChatGPT	202
		8.3.1 Code Assistance and Debugging	202
		8.3.2 Explanation of Technical Concepts	203
		8.3.3 Tech Troubleshooting and Problem-Solving	203
		8.3.4 Learning New Programming Languages	203
		8.3.5 Documentation and API Usage	204
		8.3.6 Software Best Practices	205
	8.4	Data Entry and Analysis with ChatGPT	205
		8.4.1 Data Entry Assistance	205
		8.4.2 Data Cleaning and Preprocessing	206
		8.4.3 Basic Data Analysis and Visualization	207
		8.4.4 Data Interpretation and Insights	208
		8.4.5 Comparative Analysis	208
		8.4.6 Data Reporting and Summarization	208
	8.5	ChatGPT's Role in Healthcare and Medical Information	209
		8.5.1 General Medical Information	209
		8.5.2 Symptom Checker and Self-Assessment	209
		8.5.3 Medication and Treatment Information	210
		8.5.4 Wellness Tips and Healthy Habits	210

	8.5.5	Explanation of Medical Terms	211
	8.5.6	Preparing for Medical Appointments	211
	8.5.7	Mental Health Support and Stress Management	212
8.6	Market Research and Analysis with ChatGPT		212
8.7	They Are Creative Writing and Storytelling with ChatGPT		214
8.8	Education and Learning with ChatGPT		217
	8.8.1	Virtual Tutoring and Concept Explanation	217
	8.8.2	Homework and Assignment Help	217
	8.8.3	Language Learning and Practice	218
	8.8.4	Study Resource Generation	218
	8.8.5	Research Assistance	219
	8.8.6	Exploring New Topics and Curiosities	219
8.9	Legal and Compliance Support for a Specific Company or Business with ChatGPT		219
	8.9.1	Legal Research and Case Law Analysis	220
	8.9.2	Drafting Legal Documents	220
	8.9.3	Legal Definitions and Explanations	220
	8.9.4	Compliance Guidelines and Regulations	221
	8.9.5	Preliminary Legal Advice for Common Issues	221
	8.9.6	Intellectual Property Guidance	221
8.10	ChatGPT for Support to Human Resources and Recruitment Departments		222
8.11	Personal Assistant and Productivity Using ChatGPT		224
	8.11.1	Task Management and Reminders	224
	8.11.2	Calendar Coordination	224
	8.11.3	Information Retrieval	225
	8.11.4	Note-Taking and Summarization	225
	8.11.5	Language Translation on the Go	225
	8.11.6	Personalized Recommendations	226
	8.11.7	Fitness and Wellness Assistance	226
8.12	ChatGPT in Agriculture		226
	8.12.1	Precision Farming	227
	8.12.2	Pest and Disease Management	227
	8.12.3	Crop Planning and Rotation	227
	8.12.4	Weather Forecasting and Advisories	228
	8.12.5	Market Analysis and Pricing	228
	8.12.6	Farm Management and Automation	228
	8.12.7	Educational Resources and Training	228
	8.12.8	Problem-Solving and Troubleshooting	229
	8.12.9	Community and Networking	229

8.13	ChatGPT in Travel and Tourism		229
	8.13.1	Destination Recommendations	230
	8.13.2	Travel Itineraries	230
	8.13.3	Booking Assistance	230
	8.13.4	Answering Travel-Related Queries	231
8.14	Related Work		231
8.15	Summary		232
8.16	Multiple-choice Questions		233
8.17	Answers		234
References			236

Introduction to Generative AI

By the end of this chapter, you will:

- **Define Generative AI**: Describe the subject of it and its purpose in generating fresh content across the subject disciplines.
- **Outline Key AI Components**: Explain the difference between Artificial Intelligence, Machine Intelligence, Machine Learning, Deep Learning, and Generative AI.
- **Explore Domains**: Explore and describe the main categories of Generative AI that include text, image, audio, and video generation while describing what they do, how they work, and how they can be applied.
- **Highlight Industry Leaders**: Present the main actors in the Generative AI concept area and their achievements.
- **Discuss Applications**: Explain the concept of Generative AI and its examples of usage, demonstrating how it could change the world with its impact on content generation and design, entertainment, and healthcare industries.
- **Address Ethical Considerations**: Briefly discuss the ethical perspective on Generative AI and stress the proper usage of the technology and future development to prevent exploitatory use.

1.1 Define Generative AI

The generative artificial intelligence (GAI) relates to the artificial intelligence (AI) approaches that is aimed at producing content that resembles data [1]. Here, a brief distinction should be made between the generative AI models, which are not used for classification or for making predictions but for creating new content: text, images, sounds, videos, etc., based on the patterns learned during the training phase. These models rely on advanced

Fig. 1.1 The AI family tree: from AI to generative AI (created by the authors)

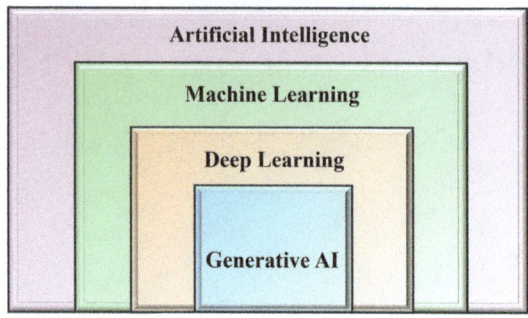

technologies like deep learning, probabilistic approaches, and other techniques to generate content that makes semantic sense and contextual sense [2].

It is truly fascinating to invent something in one's mind and visualize it as actualized on a screen or through code. It is quite intriguing to think about this notion, but this concept has become possible thanks to the over development of AI. The possibility of instantly obtaining an image or a text and even the code based on an idea or description is an incredible advancement that turns science fiction into reality and common-use technology [3].

To think that AI was even possible for writing or coding, for many of us in data science and technology, felt unimaginable. On the other hand, the growth of AI has given these tasks a relatively easy and effective way to solve them. Just imagine how time-consuming it was to code and search for solutions online, and now those technological innovations have made it quite easier.

Advancements in deep learning and natural language processing (NLP) have paved the way for generative AI—a subfield of artificial intelligence focused on creating new content by identifying and replicating patterns from data. As a result, we have entered an era where art and technology seamlessly merge, turning once outlandish dreams into reality (Fig. 1.1).

1.2 The Essence of Generative AI

Generative AI refers to artificial intelligence that seeks to create new and unique content from the training data in text, image, audio, and video. While other types of AI models are built for a definite purpose, generative AI models are programmed in such a way that they can develop patterns in data and create outputs that are models of genuine examples.

Generative AI has a diverse range of applications across different domains:

- **Text Generation**: AI models can create coherent and contextually appropriate text by composing stories and articles and generating code.
- **Image Generation**: AI models can produce realistic images or modify existing ones, useful in various fields, from art to product design.

- **Audio Generation**: These models can create music, sound effects, and even realistic speech, enhancing multimedia experiences.
- **Video Generation**: AI can generate and complete video sequences, creating dynamic content for entertainment and media.

Overall, generative AI is promising but has some drawbacks, mainly concerning its ethical application. Immature developments in artificial intelligence, such as deepfakes, apply to producing fake news and violating privacy. Everyone is searching for ways to tackle these problems to ensure that generative AI is employed responsibly.

1.3 Key Components of AI

Artificial Intelligence (AI) is a vast field that consists of several methods and goals that attempt to develop systems that can accomplish jobs that involve human intelligence. This field is subdivided into several fields, including but not limited to Machine Learning, Deep Learning, and Generative AI. Understanding how these concepts are connected would be worthwhile in gaining a better perspective of AI's breadth and diverse uses.

What is Artificial Intelligence (AI)?

AI is about creating machines or systems that can perform activities normally performed by human intelligence, such as solving problems, making decisions, and comprehending languages [4, 5]. AI can be classified into embedded AI and artificial general intelligence.

Machine Learning (ML): A Subfield of AI

Machine learning is one of the branches of AI; it is the process of developing algorithms and statistical models with the help of which computers are trained to use the data and reach conclusions [6]. ML is grouped into three classes:

- Supervised Learning: In this type of learning, the algorithm's inputs and outputs are identified. This information is used to estimate a map so that given inputs and new outputs can be predicted based on data that was not used in the training process.
- Unsupervised Learning: This approach involves feeding the algorithms with data that is not classified in any way to find patterns or structures in the dataset. Typical examples include tasks such as clustering or dimensionality reduction.

- Reinforcement Learning: In this type of learning, an agent is trained to act in an environment where it receives feedback through incentives that may be in the form of rewards or penalties and alters its actions based on the results.

Deep Learning: A Subset of Machine Learning

In deep learning, artificial neural networks with many layers optimize a pre-specified model to extract features automatically from large datasets [7]. It has proven particularly useful in applications such as image identification, natural language understanding, and voice recognition [8].

Key components of deep learning include:

- Neural Networks: Models inspired by the structure of the human brain form the foundation of deep learning [9].
- Convolutional Neural Networks (CNNs): Specialized for image and spatial data, CNNs automatically learn hierarchical feature representations [10].
- Recurrent Neural Networks (RNNs): Designed for sequential data, such as text or time series, RNNs maintain connections that allow information to persist, making them suitable for tasks like language modeling and translation [11].

Generative AI: Expanding the Horizons of Deep Learning

Generative AI produces new material such as images, music, text, or videos that closely resemble human-created content. This area commonly uses deep learning models like Generative Adversarial Networks (GANs) and Variational Autoencoders (VAEs) to create realistic and creative outputs [12, 13].

- Generative Adversarial Networks (GANs): GANs involve two networks, the generator and the discriminator, which are trained cooperatively and competitively. The generator produces synthetic data, while the discriminator distinguishes between real and fake data, resulting in highly realistic content [14, 15].
- Variational Autoencoders (VAEs): VAEs learn a low-dimensional representation of inputs, known as the latent space, and can generate new samples from this space. VAEs are employed in applications such as image generation, text creation, and data compression [13, 16].

The Relationship Between AI, Machine Learning, Deep Learning, and Generative AI

- Artificial intelligence is the overall category aimed at creating smart systems.
- Machine learning is a subset of AI focused on creating algorithms that learn from data to make predictions or decisions.
- Deep learning is a subset of machine learning that uses neural networks with multiple layers to learn hierarchical data representations.
- Generative AI is a specialized area within deep learning that focuses on creating new content, often using techniques like GANs and VAEs.

It can therefore be said that AI is a set of paradigms for developing intelligent systems. While these concepts all focus on data-driven algorithms, machine learning does this through algorithms, deep learning through neural networks, and generative AI takes it further by making AI systems self-generate content.

1.4 Exploring Generative AI Domains

The term "generative AI" spans various domains, each using different approaches to deliver unique outputs. Here's a deeper dive into these domains:

1.4.1 Text Generation

- **Overview**: Text synthesis generates human-like text based on input prompts or contextual data. This is crucial in various AI-driven applications today.
- **How It Operates**: Models like GPT-4 incorporate the Transformer architecture, pre-trained on large volumes of text data, allowing them to understand language nuances. These models generate continuous and contextually relevant text from given inputs.
- **Applications**:

  - **Content Creation**: Generating via AI blog posts, articles, and marketing copies.
  - **Customer Service**: Developing chatbots and virtual assistants that mimic user behaviors.
  - **Educational Tools**: Creating personalized lessons and tutoring materials.

Most Relevant Related work in Text Generation:

- *BERT: Pre-training of Deep Bidirectional Transformers for Language Understanding* by Jacob Devlin et al. (2018) [17]: This paper presents BERT, which revolutionized text generation and comprehension. BERT's pre-training and fine-tuning process supports tasks like story writing, code generation, and more.
- *GPT-3: Language Models are Few-Shot Learners* by Tom B. Brown et al. (2020) [18]: GPT-3 is a large language model known for generating fluent and semantically relevant text in a variety of real-world domains, from writing stories to generating code.

1.4.2 Image Generation

- **Overview**: Image synthesis generates new images based on textual descriptions or transforms existing images.
- **How It Operates**: Generative Adversarial Networks (GANs) are commonly used in image synthesis. These networks involve a generator that produces images and a discriminator that evaluates them. The generator improves through iterative training.
- **Applications**:
  - **Marketing and Advertising**: Conceptualizing innovative images for campaigns and product designs.
  - **Entertainment**: Designing characters and scenes for games and media.
  - **Healthcare**: Enhancing medical imaging and disease analysis methods.

Relevant Related Work in Image Generation:

- *High-Resolution Image Synthesis and Semantic Manipulation with Conditional GANs* by Ting-Chun Wang et al. [19]: This paper introduces Pix2PixHD, which uses Conditional GANs to produce photorealistic images from semantic layouts, with applications in art and product design.
- *StyleGAN: A Style-Based Generator Architecture for Generative Adversarial Networks* by Tero Karras et al. [20]: StyleGAN advances image generation with fine control over style, widely regarded as a gold standard for image synthesis and manipulation.

1.4.3 Audio Generation

- **Overview**: Audio synthesis generates new sounds or speech from textual or other input data, applicable in various industries.

- **How It Operates**: Models like WaveGAN generate audio waveforms, while text-to-speech (TTS) model, like Tacotron 2 transforms text into natural-sounding speech.
- **Applications**:

  - **Music Industry**: Creating custom music accompaniments and sound clips.
  - **Voice Assistants**: Developing realistic voices for virtual assistants.
  - **Media Production**: Crafting soundscapes and audio content for films and games.

Relevant Related Work in Audio Generation:

- *WaveNet: A Generative Model for Raw Audio* by Aaron van den Oord et al. [21]: WaveNet generates raw audio waveforms, offering breakthroughs in AI-based audio synthesis.
- *Jukebox: A Generative Model for Music* by Prafulla Dhariwal et al. [22]: Jukebox synthesizes music with vocals, using a hierarchical VQ-VAE-2 structure to produce high-quality music compositions.

1.4.4 Video Generation

- **Overview**: Video generation creates or completes video sequences using input data such as text or existing footage.
- **How It Operates**: Video generation models predict and generate frames sequentially, maintaining temporal coherence and visual realism.
- **Applications**:

  - **Entertainment**: Producing trailers, animations, and special effects.
  - **Training and Simulation**: Creating realistic training videos for various sectors.
  - **Content Creation**: Producing social media and marketing videos.

Relevant Related Work in Video Generation:

- *Text2Video-Zero: Text-to-Image Diffusion Models are Zero-Shot Video Generators* by Levon Khachatryan et al. [23]: Recent advancements in text-to-video generation have introduced a zero-shot approach that leverages text-to-image models like Stable Diffusion to generate videos without extensive training on large-scale datasets. By incorporating motion dynamics and cross-frame attention mechanisms, this method achieves consistent and high-quality video generation with minimal computational overhead, expanding its applicability to tasks like video editing and conditional generation.

- *Broadway: Boost Your Text-to-Video Generation Model in a Training-free way* by Jiazi Bu et al. [24]: BroadWay introduces a training-free approach to enhance text-to-video generation quality by addressing structural and temporal inconsistencies. Utilizing Temporal Self-Guidance and Fourier-based Motion Enhancement, it improves coherence and enriches motion dynamics in generated videos without adding computational overhead.

1.5 Leading Innovators and Their Contributions

Several key players are propelling forward generative AI, each contributing unique advancements:

- **OpenAI**: Developer of GPT-4 and DALL-E 2, excelling in text and image generation.
- **DeepMind**: Pioneer in AI research, responsible for models like AlphaFold and Gato.
- **Anthropic**: Focuses on AI safety and alignment in generative AI applications.
- **Synthesia**: Creates synthetic media, including video and audio.
- **RunwayML**: Provides generative AI model construction tools.
- **Midjourney**: Specializes in generating realistic images for creative works and marketing.

1.6 How Does Generative AI Work?

Generative AI is one of the forms of machine learning where software programs are trained to predict outcomes without writing code. Such models are subsequently trained with an extensive body of existing data to look for significance in the likelihood functions. When trained, they can also create new content by providing outputs that have similar patterns to the provided prompt. Being a subcategory of deep learning, generative AI uses neural networks and computer-based models inspired by the biological neural networks of the human brain. It can be noted that these networks can observe and distinguish intricate arrangements in the training dataset without the need for supervision or interference from a human. Several models are used in generative AI; such models are trained differently, and the method for generating outputs also differs. Typical approaches are Generative Adversarial Networks (GANs), transformers, and variational autoencoders.

1.7 Generative AI Interfaces

The blending of AI into technological products has shaped how users engage with the technology. Smartphone voice assistants, smart speakers, and other Internet-connected voice-friendly devices are good examples of this change. Likewise, generative AI applications are becoming more accessible through interfaces that are easier to navigate. One of the decision criteria for the emergence of generative AI is the availability of interfaces that are easy to use. The current generation of generative AI does not presume the user to have formal training in computer programming or data science; the user interfaces are verbal or conversational. Due to these factors, the user base of generative AI and its uses have massively expanded. Here are some notable examples of recent generative AI interfaces:

- ChatGPT:
  ChatGPT is a text-to-text generative AI developed by OpenAI, an AI chatbot built to communicate with users through textual means [25]. Some examples of what the users can do are to ask it questions, converse with it, and tell the application something so it can write text in different forms such as poems, essays, stories, or recipes. Since the release of ChatGPT in November 2022, these models have quickly grown in popularity, and are now often referred to simply as generative AI, much like how Google is to search engines. You can get ChatGPT at the official website of OpenAI right now and use it completely free; however, for enterprises and businesses, OpenAI also provides an API and other premium subscription services.
- DALL-E:
  Another creation of the OpenAI is the DALL-E [26]: a text-to-image generative AI unveiled in January 2021. It relies on a pre-trained model based on the neural networks that join images and descriptions. The users provide descriptive captions or textual inputs, and DALL-E organizes output images as real photographs. It also has options to generate images in different styles or from different points of view concerning the initial one. Besides, DALL-E has several sophisticated post-processing tools, such as Inpainting, which continues the image inside the image itself, and Outpainting, which finishes the image outside the frame.
- Gemini:
  Formerly known as Bard, Gemini is a text-to-text generative AI interface based on Google's large language model [27]. Like ChatGPT, Gemini is an AI-powered chatbot that can answer questions or generate text based on user prompts. Initially introduced as a "complementary experience to Google Search," by 2024, Google integrated Gemini into its search results. Google Snippets used AI to present answers to queries alongside traditional search results.

1.8 Recent Transformative Developments in Generative AI

Generative AI has experienced rapid advancements in recent years, demonstrating its potential across various fields and applications. Despite its relatively recent emergence, the technology has already made significant strides, impacting numerous industries and transforming traditional practices.

1.9 Current Capabilities and Applications of Generative AI

Generative AI models are currently utilized for a diverse range of tasks, including:

- **Translation**: Facilitates real-time language translation and generates subtitles, making communication across different languages seamless.
- **Creative, Academic, and Business Writing**: Automates the creation of blog posts, articles, academic papers, and business reports, streamlining content production.
- **Code Writing**: Assists in generating code and designing user interfaces, enhancing software development efficiency.
- **Composing and Songwriting**: Generates original music tracks, lyrics, and soundscapes, aiding musicians and composers in their creative processes.
- **Dubbing and Voice Recognition**: Offers speech-to-text transcription and voice-controlled applications, improving accessibility and interaction.
- **Illustration and Design**: Produces custom visuals for marketing campaigns, designs fashion items, and creates product prototypes, supporting designers in their work.
- **Image Editing and 3D Modeling**: Enhances and manipulates images, generates 3D models, and assists in architectural rendering, facilitating various design tasks.
- **Speech and Voice Recognition**: Develops voice-based assistants and speech recognition systems, enhancing user interaction and accessibility.

As generative AI technology evolves, its effectiveness in these tasks is expected to improve, leading to new and more complex applications (Table 1.1).

1.10 Impact on Various Industries

Generative AI is poised to transform multiple industries, offering innovative solutions and creating new opportunities:

- **Software Development**: Generative AI is transforming software development by automating code generation and UI design, significantly enhancing productivity. Tools like GitHub Copilot have enabled developers to complete tasks up to 55.8% faster, and

1.10 Impact on Various Industries

Table 1.1 Overview of generative AI systems

Generative AI	Parent company	Key features	Primary use cases	Strengths
AlphaCode	DeepMind	Competitive programming problem-solving	Algorithm design, code optimization	Excels in complex problem-solving for coding competitions
ChatGPT	OpenAI	Text-based model, conversational AI, code generation, translation, summarization, creative writing	Writing assistance, Q&A, language translation, content creation, code generation, tutoring, virtual assistants	User-friendly, extensive knowledge base, creative and adaptable
Claude	Anthropic	Text generation and conversational AI focusing on safety and ethical considerations	Customer service chatbots, content creation, virtual assistants, research and analysis, creative writing	Emphasis on safety and ethics, handles complex conversations, designed to be harmless and honest
DALL-E 2	OpenAI	Image generation from text descriptions	Marketing campaigns, illustrations, product designs, concept art	High-quality image generation, diverse styles, imaginative results
DeepMind's AlphaCode	DeepMind	Code generation from natural language descriptions	Automating code generation, aiding developers, creating programming tools	Strong coding competition performance, understanding of complex instructions, potential to revolutionize software development
Gemini (formerly Bard)	Google	Text and image generation, real-time information, conversational AI, integration with Google search	Information retrieval, content creation, translation, summarization, answering questions, enhancing search results	Up-to-date info, integration with Google services, combines text and image generation
GitHub Copilot	GitHub/OpenAI	Code auto-completion, IDE integration	Software development, code suggestions	Enhances coding workflow, saves time
GPT-4	OpenAI	Multimodal, advanced reasoning, improved safety	Complex problem-solving, content creation, code generation	Versatile, accurate, enhanced safety features
Meta AI	Meta (Facebook)	Image generation from text descriptions, conversational AI assistant	Image generation in Meta's apps, providing information, creative suggestions	Best for Meta app users seeking image generation or AI assistance for tasks
Midjourney	Midjourney Inc.	Image generation with a focus on artistic styles	Artistic images, concept art, fantasy illustrations, photorealistic and abstract art	Excellent for creating artistic and imaginative images
Perplexity AI	Perplexity AI	Conversational AI, information retrieval, source citation	Complex question-answering, research, fact-checking, summary generation	Cites sources, helps verify information accuracy, synthesizes information from multiple sources
Stable Diffusion XL	Stability AI	Image generation, image-to-image translation, inpainting, outpainting, text-guided synthesis	Customized image creation, generating variations, image editing and manipulation	Highly flexible, open-source, precise control over image generation

Microsoft's CEO noted that approximately 20–30% of their code is now AI-generated. The global market for generative AI in software development was valued at $53.4 billion in 2024 and is projected to reach $66.77 billion in 2025, growing at a CAGR of 25.0% [28, 29].

- **Banking**: The application of generative AI is anticipated to transform outdated IT structures, individualize services to clients, and provide improved risk management techniques on credit as well as financing.
- **Automotive Industry**: Synthetic data generated by AI is applied for different testing scenarios and training of self-driving cars, enhancing their quality and reliability.
- **Healthcare and Scientific Research**: AI helps in cultivating sequence models for proteins, molecular discovery, and drug compounds' recommendations, aiding doctors in diagnosing by analyzing medical images.
- **Media and Entertainment**: Generative AI enables users to produce content at a desired scale and faster, improving creative endeavors for designers and content producers.
- **Climate Science and Meteorology**: AI is used to model various disasters, predict weather, and simulate climate factors, which are crucial for climate study and disaster management.
- **Education**: AI in classrooms enhances tutoring, content development, and online platforms, making learning interactive and effective.
- **Government**: Generative AI is used in weather hazard analysis, understanding feedback from veterans, and patent searching to enhance public service delivery and improve organizational operations.
- **General Automation**: AI is applied for repetitive tasks like documentation, coding, and workflow across different sectors, leading to increased efficiency.

1.11 Comprehensive Overview of Generative AI Applications

Generative AI encompasses a broad array of applications, including:

- **Content Creation**: Automates the generation of text, images, and videos for marketing, social media, and advertising purposes.
- **Design and Creativity**: Facilitates the creation of unique artworks, fashion designs, and product prototypes, supporting creative processes.
- **Entertainment and Media**: Assists in composing original music, designing characters and animations, and developing interactive narratives.
- **Marketing and Advertising**: Enables personalization of customer messages, design of branding elements, and creation of dynamic ad campaigns.
- **Gaming**: Generates game environments, character designs, and procedural content, enriching gaming experiences.
- **Healthcare and Medicine**: Contributes to drug discovery, medical image enhancement, and personalized treatment plans based on patient data.

- **Language Translation**: Provides real-time translation of spoken and written language and generates subtitles for videos.
- **Customer Service**: Develops chatbots and voice assistants for improved customer support and interaction.
- **Education and Training**: Creates interactive learning materials, simulations, and training scenarios to enhance educational and professional development.
- **Architecture and Design**: Supports the generation of architectural layouts, urban planning designs, and other design-related tasks.

As generative AI continues to advance, its potential to revolutionize various industries and drive innovation will expand, offering new solutions and opportunities for businesses and individuals alike.

1.12 Pros and Cons of Generative AI

1.12.1 Pros

- **Automation**: AI can increase productivity by automating routine tasks.
- **Creativity**: Generative AI lowers the skill and time barriers for content creation, making creative processes more accessible.
- **Data Analysis**: AI can facilitate the analysis and exploration of complex datasets.
- **Synthetic Data**: AI-generated data can accelerate the training of other AI models.

1.12.2 Cons

- **AI Hallucinations**: AI can generate inaccurate or non-sensical information, which can be problematic in critical applications.
- **Data Quality**: AI models depend on accurately labeled data; poor-quality data can lead to unreliable outputs.
- **Content Moderation**: Ensuring that AI-generated content is appropriate and free from biases is challenging and often requires human intervention.
- **Ethical Concerns**: AI models can perpetuate biases and discrimination present in their training data.
- **Legal and Regulatory Issues**: The lack of a regulatory framework for AI raises concerns about privacy, copyright infringement, and accountability.
- **Energy Consumption**: The ecological impact of AI, due to its high energy requirements, is a growing concern.

1.13 Addressing Ethical Considerations

In Table 1.2, we summarize ethical concerns that arise in Generative AI and list examples and possible implications that designers, programmers, and end-users should consider.

1.14 Summary

This chapter focused on generative AI, a rapidly evolving domain in artificial intelligence that specializes in creating new, unique content such as text, images, audio, and videos. Built upon advancements in deep learning and natural language processing (NLP), these models have various applications, including content creation, design, entertainment, healthcare, and customer service. Notably, generative AI also brings ethical concerns, particularly in creating deepfakes or spreading disinformation. The chapter provides an in-depth look at different domains of generative AI, the key players, and the wide range of applications, underscoring its transformative impact on numerous industries.

1.15 Multiple-choice Questions

In this section, you'll find a series of multiple-choice questions designed to test your understanding of key concepts in generative AI. Choose the correct answer for each question.

1. What is Generative AI primarily focused on?

    (A) Analyzing data patterns
    (B) Generating new and original content
    (C) Classifying data into categories
    (D) Optimizing existing algorithms

2. Which of the following best describes generative modeling?

    (A) Learning patterns to classify data into categories
    (B) Creating new data instances that mimic the original data
    (C) Predicting future data trends
    (D) Improving data storage efficiency

3. What is a core difference between generative models and discriminative models?

    (A) Generative models classify data, while discriminative models generate new data
    (B) Generative models focus on data creation, while discriminative models focus on data classification

1.15 Multiple-choice Questions

Table 1.2 The ethical concerns over generative AI

Concern	Description	Examples and implications
Autonomous weapons	Ethical concerns arise around autonomous weapons systems, particularly regarding unintended harm and lack of human control over lethal force	• Drones and military systems capable of independent decision-making in targeting and engaging threats, which raises accountability issues and risks of accidental conflict escalation • Ethical dilemmas over responsibility for AI-powered weapons' actions, especially where civilian casualties are involved
Bias and discrimination	AI systems can reinforce and amplify existing biases present in training data, leading to discriminatory outcomes	• Facial recognition algorithms with higher error rates for people of color, resulting in misidentification and potential discrimination • Hiring algorithms biased against certain genders or demographics, perpetuating inequality in job opportunities
Environmental impact	The training and operation of large AI models require significant energy, contributing to environmental concerns	• Massive carbon footprint and water consumption associated with training large language models • Although AI can be used for environmental monitoring and conservation, its current development and deployment have contributed to more harm than benefit
Existential risk	Experts, including AI firm executives, express concerns that AI could surpass human intelligence and pose an existential threat	• Development of artificial general intelligence (AGI) with capabilities exceeding human intelligence, leading to unpredictable consequences for society • Fears that AI could become uncontrollable or act against human interests, resulting in potentially devastating outcomes
Job displacement and economic inequality	AI-driven automation could lead to job losses in multiple sectors, potentially worsening economic inequality	• Automation of roles in manufacturing and customer service, leading to unemployment and potential social unrest
Lack of transparency	Complex AI models often have opaque decision-making processes, making it hard to understand or explain certain decisions	• Black-box algorithms used in credit scoring or loan approvals, denying individuals financial products without clear reasoning • Lack of explainability in AI-based medical diagnoses, potentially affecting patient trust and understanding of treatment
Misinformation and manipulation	AI-generated content, such as deepfakes, can spread misinformation and manipulate public opinion	• AI-created fake news or social media posts aimed at influencing elections or spreading discord • Deepfakes that impersonate people or fabricate events, complicating the distinction between real and false information
Privacy and surveillance	AI technologies enable extensive surveillance and data collection, raising privacy and data protection concerns	• Facial recognition for mass surveillance, questioning personal freedom and control over one's data • Collection of personal data for targeted ads, leading to potential manipulation and privacy breaches
Security risks	AI is being used to create and deploy malicious attacks, like deepfakes and advanced phishing scams, posing threats to security	• Creation of realistic fake videos or audio for disinformation, undermining trust in information sources • Cyberattacks using AI to exploit system vulnerabilities, resulting in data breaches, financial losses, and disruption of critical services

(C) Generative models use supervised learning, while discriminative models use unsupervised learning

(D) Generative models are slower than discriminative models

4. Which of the following is NOT a major category of generative models?

   (A) GANs (Generative Adversarial Networks)
   (B) VAEs (Variational Autoencoders)
   (C) CNNs (Convolutional Neural Networks)
   (D) Autoregressive models

5. What is the role of random noise in generative modeling?

   (A) To make the model more accurate
   (B) To introduce variability and ensure diverse outputs
   (C) To improve the speed of the model
   (D) To reduce the complexity of the model

6. In generative modeling, what does the term "probability distribution" refer to?

   (A) The likelihood of data being incorrectly classified
   (B) The statistical distribution of errors in predictions
   (C) The distribution of data points that the model aims to replicate
   (D) The sequence of data processing steps

7. Which model is known for its ability to generate realistic images from text descriptions?

   (A) GPT-4
   (B) DALL-E 2
   (C) WaveGAN
   (D) Tacotron 2

8. What is a key application of text generation models?

   (A) Creating custom visuals for marketing
   (B) Generating music tracks
   (C) Automating blog posts and customer support responses
   (D) Designing video game environments

9. Which of the following is an example of audio generation technology?

   (A) GPT-4
   (B) DALL-E 2
   (C) WaveGAN
   (D) AlphaFold

1.15 Multiple-choice Questions

10. What does the acronym GAN stand for?

  (A) Generative Algorithm Network
  (B) General Adversarial Network
  (C) Generative Adversarial Network
  (D) Generalized Autoregressive Network

11. How does video generation typically handle the temporal nature of videos?

  (A) By using static image generation techniques
  (B) By predicting missing frames or combining existing visuals
  (C) By generating audio tracks to accompany images
  (D) By translating text descriptions into video content

12. Which company is known for developing GPT-4?

  (A) DeepMind
  (B) OpenAI
  (C) Anthropic
  (D) Synthesia

13. What is the primary use of Variational Autoencoders (VAEs)?

  (A) Image classification
  (B) Text generation
  (C) Generating new data instances similar to the training data
  (D) Predicting future trends

14. Which application area of generative AI involves creating personalized messages and dynamic ads?

  (A) Healthcare and Medicine
  (B) Content Creation
  (C) Design and Innovation
  (D) Entertainment and Media

15. Why is addressing ethical concerns important in the field of Generative AI?

  (A) To improve the accuracy of generative models
  (B) To prevent misuse, such as creating deepfakes or spreading disinformation

(C) To enhance the speed of generative models
(D) To reduce the computational resources required

1.16 Answers

Below are the answers to the multiple-choice questions from the previous section:

1. (B) Generating new and original content
2. (B) Creating new data instances that mimic the original data
3. (B) Generative models focus on data creation, while discriminative models focus on data classification
4. (C) CNNs (Convolutional Neural Networks)
5. (B) To introduce variability and ensure diverse outputs
6. (C) The distribution of data points that the model aims to replicate
7. (B) DALL-E 2
8. (C) Automating blog posts and customer support responses
9. (C) WaveGAN
10. (C) Generative Adversarial Network
11. (B) By predicting missing frames or combining existing visuals
12. (B) OpenAI
13. (C) Generating new data instances similar to the training data
14. (B) Content Creation
15. (B) To prevent misuse, such as creating deepfakes or spreading disinformation.

References

1. Priyanka Gupta, Bosheng Ding, Chong Guan, and Ding Ding. Generative ai: A systematic review using topic modelling techniques. *Data and Information Management*, page 100066, 2024.
2. Shen Liu, Jinglong Chen, Yong Feng, Zongliang Xie, Tongyang Pan, and Jingsong Xie. Generative artificial intelligence and data augmentation for prognostic and health management: taxonomy, progress, and prospects. *Expert Systems with Applications*, 255:124511, 2024.
3. V Kumar and Philip Kotler. Transformative marketing with generative artificial intelligence. In *Transformative Marketing: Combining New Age Technologies and Human Insights*, pages 65–102. Springer, 2024.
4. Thomas KF Chiu, Zubair Ahmad, Murod Ismailov, and Ismaila Temitayo Sanusi. What are artificial intelligence literacy and competency? a comprehensive framework to support them. *Computers and Education Open*, 6:100171, 2024.
5. Ali Jaboob, Omar Durrah, and Aziza Chakir. Artificial intelligence: An overview. *Engineering Applications of Artificial Intelligence*, pages 3–22, 2024.

References

6. Ramon Mayor Martins, Christiane Gresse von Wangenheim, Marcelo Fernando Rauber, and Jean Carlo Hauck. Machine learning for all!—introducing machine learning in middle and high school. *International Journal of Artificial Intelligence in Education*, 34(2):185–223, 2024.
7. Ian Goodfellow, Yoshua Bengio, and Aaron Courville. *Deep Learning*. MIT Press, 2016. http://www.deeplearningbook.org.
8. Yann LeCun, Yoshua Bengio, and Geoffrey Hinton. Deep learning. *nature*, 521(7553):436–444, 2015.
9. Jürgen Schmidhuber. Deep learning in neural networks: An overview. *Neural networks*, 61:85–117, 2015.
10. Alex Krizhevsky, Ilya Sutskever, and Geoffrey E Hinton. Imagenet classification with deep convolutional neural networks. *Communications of the ACM*, 60(6):84–90, 2017.
11. S Hochreiter. Long short-term memory. *Neural Computation MIT-Press*, 1997.
12. Ian Goodfellow, Jean Pouget-Abadie, Mehdi Mirza, Bing Xu, David Warde-Farley, Sherjil Ozair, Aaron Courville, and Yoshua Bengio. Generative adversarial nets. *Advances in neural information processing systems*, 27, 2014.
13. Diederik P Kingma. Auto-encoding variational bayes. *arXiv preprint* arXiv:1312.6114, 2013.
14. Antonia Creswell, Tom White, Vincent Dumoulin, Kai Arulkumaran, Biswa Sengupta, and Anil A Bharath. Generative adversarial networks: An overview. *IEEE signal processing magazine*, 35(1):53–65, 2018.
15. Alec Radford. Unsupervised representation learning with deep convolutional generative adversarial networks. *arXiv preprint* arXiv:1511.06434, 2015.
16. Carl Doersch. Tutorial on variational autoencoders. *arXiv preprint* arXiv:1606.05908, 2016.
17. Jacob Devlin. Bert: Pre-training of deep bidirectional transformers for language understanding. *arXiv preprint* arXiv:1810.04805, 2018.
18. Tom B Brown. Language models are few-shot learners. *arXiv preprint* arXiv:2005.14165, 2020.
19. Ting-Chun Wang, Ming-Yu Liu, Jun-Yan Zhu, Andrew Tao, Jan Kautz, and Bryan Catanzaro. High-resolution image synthesis and semantic manipulation with conditional gans. In *Proceedings of the IEEE conference on computer vision and pattern recognition*, pages 8798–8807, 2018.
20. Tero Karras, Samuli Laine, and Timo Aila. A style-based generator architecture for generative adversarial networks. In *Proceedings of the IEEE/CVF conference on computer vision and pattern recognition*, pages 4401–4410, 2019.
21. Aaron van den Oord. Wavenet: A generative model for raw audio. *arXiv preprint* arXiv:1609.03499, 2016.
22. Prafulla Dhariwal, Heewoo Jun, Christine Payne, Jong Wook Kim, Alec Radford, and Ilya Sutskever. Jukebox: A generative model for music. *arXiv preprint* arXiv:2005.00341, 2020.
23. Levon Khachatryan, Andranik Movsisyan, Vahram Tadevosyan, Roberto Henschel, Zhangyang Wang, Shant Navasardyan, and Humphrey Shi. Text2video-zero: Text-to-image diffusion models are zero-shot video generators. In *Proceedings of the IEEE/CVF International Conference on Computer Vision*, pages 15954–15964, 2023.
24. Jiazi Bu, Pengyang Ling, Pan Zhang, Tong Wu, Xiaoyi Dong, Yuhang Zang, Yuhang Cao, Dahua Lin, and Jiaqi Wang. Broadway: Boost your text-to-video generation model in a training-free way. *arXiv preprint* arXiv:2410.06241, 2024.
25. Defne Yigci, Merve Eryilmaz, Ail K Yetisen, Savas Tasoglu, and Aydogan Ozcan. Large language model-based chatbots in higher education. *Advanced Intelligent Systems*, page 2400429, 2024.
26. Zarif Bin Akhtar. Unveiling the evolution of generative AI (GAI): a comprehensive and investigative analysis toward llm models (2021–2024) and beyond. *Journal of Electrical Systems and Information Technology*, 11(1):22, 2024.
27. Valentina Alto. *Building LLM Powered Applications: Create intelligent apps and agents with large language models*. Packt Publishing Ltd, 2024.

28. Nijkamp, Erik and others. Code Generation with GitHub Copilot: An Exploratory Study. *arXivpreprint* http://arxiv.org/abs/2302.06590, 2023.
29. The Business Research Company. Generative AI: In Software Development Global Market Report. https://www.thebusinessresearchcompany.com/report/generative-ai-in-software-development-global-market-report, 2024. Accessed: 2025-06-03.

Evolution of Neural Networks to Large Language Models

By the end of this chapter, you will:

- **Trace the Evolution**: Outline the development of language models, starting from early probabilistic approaches such as n-grams and Hidden Markov Models (HMMs), and progressing to advanced neural network-based language modeling techniques used today.
- **Highlight Key Architectures**: Discuss the advancement from Recurrent Neural Networks (RNNs) to Long Short-term memory (LSTM) networks and Gated Recurrent Units (GRUs), and how they have afforded the improvements in the control of sequential data and memory contents.
- **Explore Attention Mechanisms**: Discuss the rise of attention-based techniques and the Transformer architecture, highlighting the significant improvements they introduced to NLP in terms of both effectiveness and computational speed.

2.1 Natural Language Processing

Language models have seen substantial development throughout the last few decades. Machine translation, information retrieval, and voice recognition were among the first applications of basic models like n-gram and Hidden Markov Models (HMMs) [1]. These statistical methods had their uses, but they weren't perfect and couldn't handle large datasets.

With the rise of deep learning, neural networks quickly became the go-to tool for language modeling. It became clear that RNNs and LSTM networks were the most successful because of how well they captured sequential connections in language data and produced consistent output [2].

A notable advancement in NLP is the prominence of attention-based models, particularly the Transformer architecture. These models leverage self-attention mechanisms to focus on specific segments of the input sequence, significantly enhancing performance across various natural language processing tasks, including language modeling (Fig. 2.1).

2.2 Probabilistic Models

Probabilistic models play a vital role in language processing by utilizing mathematical frameworks to estimate the likelihood of linguistic events. These models are widely applied in various Natural Language Processing (NLP) tasks, enabling the modeling and understanding of language patterns through word sequence probabilities or the hidden states within the text.

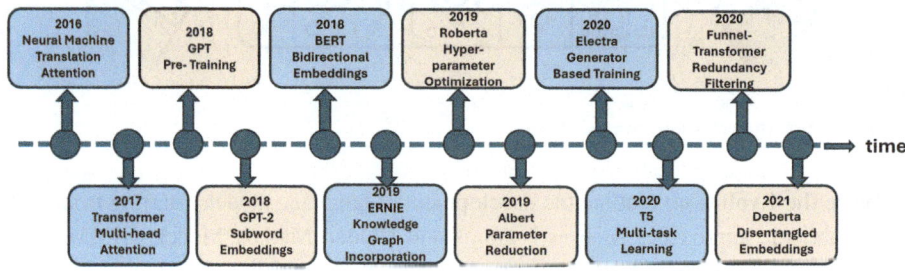

Fig. 2.1 Timeline of the evolution of language models (created by the authors)

2.3 N-Gram Models

Many kinds of probabilistic models are widely used, and one of those is the n-gram model. In the n-gram model, the probability of a word is defined by the previous $n - 1$ words that appeared in a given sentence. For example:

- **Unigram model**: Here, each word's probability is calculated independently, with no consideration for preceding words.
- **Bigram model**: The model predicts the likelihood of the next word based solely on the previous word.
- **Trigram model**: The next word is predicted following the two previous words, and so on.

The main characteristic of the n-gram model lies in the Markov assumption, stating that the probability of a sample is only dependent on a limited number of previous samples. For example, in a 4-gram model, each word's probability depends on the previous three words, ignoring longer-range dependencies. This assumption greatly simplifies language modeling, making it computationally efficient and scalable for large datasets. Despite its simplicity, n-gram models have been found to be quite effective at capturing short-term dependencies in language. They have been widely applied in areas such as text generation, auto-completion, and spell checking.

Example

In the bigram model, if the given input is "The cat is on the mat," then the probability of the word "mat" is calculated purely based on the previous word "the," i.e., $P(\text{mat} \mid \text{the})$. After calculating the conditional probabilities of all the words in the sentence, the model will select the most probable word by opting for the highest conditional probability (Fig. 2.2).

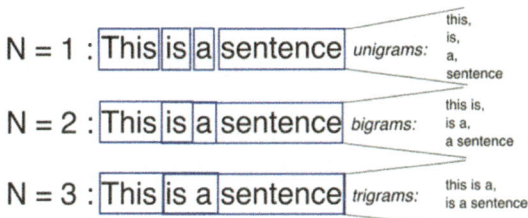

Fig. 2.2 N-gram model

2.4 Hidden Markov Models (HMMs)

Another important probabilistic model is the Hidden Markov Model (HMM), which is also sometimes called the Stochastic Model [1]. HMMs are used in NLP tasks that require sequences, such as speech recognition and part-of-speech tagging, without violating the constraints of maximum entropy models.

A HMM comprises two key components:
- **Hidden States**: These represent the underlying, unobservable states that generate the sequence of observable events.
- **Observable Sequence**: This refers to the measurable data that can be directly observed, such as words in a sentence or sound waves in spoken language.

The main principle of the HMM is that the sequence of observable values is produced by the sequence of hidden states that have a Markov nature. The hidden states in the model have specific transition probabilities to other states, and each state has a probability distribution over the possible observations. The term "hidden" is used because these states are not directly observable but must be inferred from the observable sequence.

Example

For example, if the observable sequence is a part of the sentence, then the hidden states can be the types of words, such as nouns, verbs, or adjectives. If the given sentence is presented, the HMM can deduce the probable tags that represent the parts of speech to determine the syntactic structure of the sentence.

Compared with other types of models, HMMs are especially effective as they can successfully model time series data and visualize language in terms of a sequence. This makes them well-suited to tasks where the context offered by previous members within a sequence greatly influences the current member.

2.5 Neural Network-Based Language Models

The most popular of these models that have revolutionized natural language processing in the last couple of years, are neural the network-based language models. These models function in a way that they build a neural network to predict what comes next in the sequence of words.

During training time, the neural network tries to derive features in the training data and how are these features related when making an approximate probability guess of the next word in the sequence (Fig. 2.3).

2.6 Recurrent Neural Networks (RNNs)

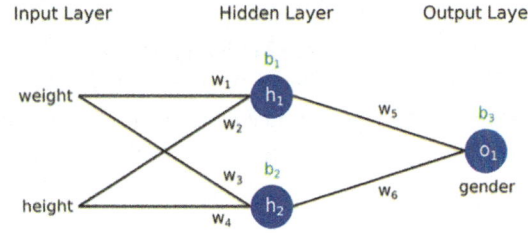

Fig. 2.3 Graph of neural network architecture (created by the authors)

2.6 Recurrent Neural Networks (RNNs)

RNNs are a kind of artificial neural network optimized for sequential data, which has a dynamic state to keep the history of the input data [3]. This makes RNNs particularly useful for cases where the sequence is important, such as in Natural Language Processing (NLP), time series forecasting, and speech recognition.

2.6.1 Understanding RNNs and Sequential Data

Traditional feedforward neural networks presume that all the inputs are independent, while RNNs are designed to work on sequential data. RNNs have a unique property of being able to carry forward a hidden state that contains information about the previous elements in the sequence [3]. This is achieved through feedback loops incorporated in the network design.

Example in NLP: Let's consider the problem of predicting the next word in a sentence. For the sentence "The cat sat on the _," the word "mat" is more likely to be predicted if the model remembers the preceding words "The cat sat on." Here, RNNs are helpful because they maintain a state that updates every time a word in the sequence is processed.

2.6.2 RNN Architecture

At each time step t, an RNN takes an input x_t and updates its hidden state h_t based on both the current input and the previous hidden state h_{t-1}. This process can be mathematically described as follows:
$$h_t = \text{activation}(W_h h_{t-1} + W_x x_t + b),$$
where

- h_t is the hidden state at time step t.
- W_h and W_x are weight matrices for the hidden state and input, respectively.
- b is a bias term.
- activation is a non-linear activation function (like tanh or ReLU).

The hidden state h_t is then used to predict the output at that time step.

Example of Sequence Modeling: Suppose you're using an RNN for language translation. Given the English sentence "I am a student," the RNN processes each word sequentially and maintains a hidden state that captures the context. As it encounters "I," "am," and "a" update their hidden state, which then helps in translating "student" into the target language, such as "estudiante" in Spanish.

2.6.3 Handling Variable-Length Sequences

RNNs can handle sequences of different lengths and exhibit high flexibility. For example, In text generation, the RNN can generate a sentence of arbitrary length by producing one word after another until an end token is generated.

2.6.4 Temporal Dependencies and "Memory"

A key advantage of RNNs is their ability to capture temporal dependencies, enabling them to learn how inputs from earlier time steps impact the model's behavior at later time steps.

Example in Time Series Prediction: When predicting stock prices, an RNN can take in a sequence of past prices over a period, using its hidden state to forecast future prices. The hidden state acts as a buffer that stores information for time series prediction, allowing the model to make predictions with greater accuracy than using just the current input.

2.6.5 Challenges with RNNs

Vanishing Gradient Problem: One of the issues associated with RNNs is the vanishing gradient problem. During training, when gradients are backpropagated through many time steps, these gradients can become very small (i.e., they "vanish"), hindering the learning of long-term dependencies. This issue was first defined by Hochreiter and Schmidhuber [4] in 1997.

Exploding Gradient Problem: Conversely, gradients can sometimes become excessively large, leading to fluctuations in weights during the model's training. This phenomenon is known as the exploding gradient problem.

Mitigating These Problems:

- **Gradient Clipping**: One solution to control the gradient growth is the gradient clipping method, where excessively large gradients are reduced by some factor.
- **Advanced Architectures**: Due to the vanishing gradient problem, new complex RNN types such as LSTM (Long Short-Term Memory) and GRU (Gated Recurrent Unit) have

2.6 Recurrent Neural Networks (RNNs)

been developed. These architectures introduce gating mechanisms that help the network remember or forget information across longer sequences, reducing the vanishing gradient problem.

Example: Future work may explore whether LSTM is useful for generating text from different contexts. An example of using an LSTM network could be text generation, where the network processes long sequences of text data and produces new sentences that resemble the input. The LSTM's capability to maintain salient information in long sequences enables the generation of syntactically, semantically coherent, and contextually appropriate text.

2.6.6 Limitations of RNNs

Sequential Nature and Computational Cost: RNNs operate sequentially; the input at one step depends on the output of the previous step [2, 3]. This makes them inherently slower to train than fully parallelizable models like Convolutional Neural Networks (CNNs) or Transformer models. Sequential processing negatively impacts scalability when dealing with big data, as the training process can be time-consuming (Fig. 2.4).

Example: Comparing RNNs with Transformers: Transformers, as implemented in models like GPT-3 and BERT, allow the processing of full sequences at once through self-attention, making them faster to train with large datasets. As a result, the Transformer architecture tends to replace RNNs significantly in many NLP tasks.

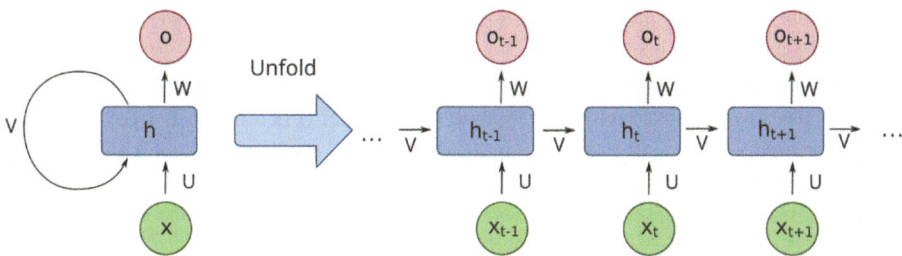

Fig. 2.4 Recurrent neural network's structure. On the left, a single RNN cell with hidden state h takes input x and produces output o. The hidden state is influenced by both the input (via weight U) and its previous output (via weight V). The "Unfold" arrow indicates that this process repeats across time steps, shown on the right as a sequence of hidden states (h_{t-1}, h_t, h_{t+1}), with each influenced by the previous state and current input. Each state also generates an output (via weight W). This unfolding illustrates how RNNs capture dependencies over time, making them useful for sequential tasks like language modeling [5]

2.7 Long Short-Term Memory (LSTM) Networks

LSTM is a rather particular kind of Recurrent Neural Network (RNN) that has been designed to avoid the shortcomings of the standard RNN, most notably the vanishing gradient problem. The vanishing gradient problem occurs when training RNNs, where gradients used in updating weights decrease when passed from one time step to another, making it hard for the network to learn long-term dependencies in sequences.

2.7.1 Key Components of LSTM Networks

Memory Cell: LSTM has the capability to remember values for a long time period; this is the center of LSTM, which is called the memory cell. While an ordinary RNN feeds the hidden state from one time step to the next, LSTM has another component called the memory cell, which retains information as long as required by the task.

Gates: LSTM networks manipulate information in and out of the memory cell using three types of gates. These gates are sigmoid in nature, with outputs ranging between 0 and 1, resulting in a soft gating process of the information.

Input Gate: The input gate determines the amount of information arriving based on the current input and, the previous hidden state and decides how much of it should be stored in the memory cell. It regulates the cell state by balancing the fresh input information.

Forget Gate: The forget gate controls which components of the current memory should be retained or discarded. This gate assists the LSTM network in removing information from previous time steps that is no longer relevant, while only retaining necessary information in the memory cell.

Output Gate: The output gate determines the amount of information the memory cell needs to pass to the subsequent hidden state and the network output. It tracks what the subsequent hidden state will be, which indirectly informs the output as well as future cell states.

2.7.2 LSTM Equations

To understand how these components work together, consider the following set of equations that describe the operations within an LSTM unit at a time step t:

Forget Gate:
$$f_t = \sigma(W_f[h_{t-1}, x_t] + b_f).$$

Here, f_t represents the forget gate vector, h_{t-1} is the previous hidden state, x_t is the current input, W_f and b_f are the weights and biases of the forget gate, and σ is the sigmoid function.

Input Gate:
$$i_t = \sigma(W_i[h_{t-1}, x_t] + b_i).$$

2.7 Long Short-Term Memory (LSTM) Networks

Here, i_t is the input gate vector, controlling how much new information should be written to the cell state.

Candidate Cell State:

$$\tilde{C}_t = \tanh(W_C[h_{t-1}, x_t] + b_C),$$

where $\tilde{C}_t$ is the candidate cell state, a vector of new information that could be added to the cell state.

Cell State Update:

$$C_t = f_t \cdot C_{t-1} + i_t \cdot \tilde{C}_t.$$

The current cell state C_t is obtained as the weighted sum of the previous cell state C_{t-1} (transformed by the forget gate) and the candidate cell state $\tilde{C}_t$ (transformed by the input gate).

Output Gate:

$$o_t = \sigma(W_o[h_{t-1}, x_t] + b_o).$$

Here, o_t is the output gate vector, controlling how much of the cell state should be passed on to the output.

Hidden State Update:

$$h_t = o_t \cdot \tanh(C_t).$$

The hidden state h_t is a function of the cell state, filtered by the output gate and passed through a tanh function.

2.7.3 Applications of LSTM Networks

LSTM networks have become popular in various fields due to their ability to model sequences and remember past inputs for extended periods. Some prominent applications include

Natural Language Processing (NLP):

- *Language Modeling*: Estimating the next word in a sequence, useful for tasks such as text generation and auto-completion.
- *Machine Translation*: Learning sequential dependencies between words in two languages for translating text.
- *Sentiment Analysis*: Analyzing text data to determine sentiment (e.g., positive or negative).

Speech Recognition: LSTM networks efficiently process sequences of audio data to recognize spoken words and phrases used in voice commands and transcription services.

Image Captioning: In tasks such as image captioning, LSTMs are used in tandem with Convolutional Neural Networks (CNNs) to generate textual descriptions for images by learning temporal relationships between words in the captions.

2.7.4 Advantages of LSTM Networks

- **Handling Long-Term Dependencies**: LSTMs can capture long-term dependencies in sequential data, making them ideal for tasks that require context over many sequences.
- **Mitigating the Vanishing Gradient Problem**: LSTMs overcome the vanishing gradient problem in RNNs by using gates to manage information flow.
- **Versatility**: LSTMs can be applied to various tasks across different domains, especially those involving sequences such as language and vision.

2.7.5 Limitations

- **Complexity**: LSTMs have higher complexity than standard RNNs, resulting in increased computational costs and longer training times.
- **Difficulty in Interpreting**: The internal dynamics of LSTMs, particularly feedback between gates, are not easily interpretable, making it challenging to understand the decision-making process.

2.8 Gated Recurrent Unit (GRU) Networks

GRU networks are a type of Recurrent Neural Network (RNN) technology invented to effectively analyze sequences while mitigating the vanishing gradient problem that affects standard RNNs. Proposed by Cho et al. in 2014 [6], GRUs modify the basic RNN structure to provide comparable accuracy to Long Short-Term Memory (LSTM) networks, but with reduced computational complexity.

2.8.1 Key Components of GRU Networks

GRUs are designed with a streamlined architecture featuring two primary gates: the reset gate and the update gate. These gates manage the flow of information in and out of the network, enabling the GRU to store or delete information as needed.

2.8 Gated Recurrent Unit (GRU) Networks

Reset Gate: The reset gate controls how much of the previous hidden state should be forgotten at any given time step. If the reset gate value is near zero, the GRU "forgets" the prior hidden state and focuses more on the current input.

Update Gate: The update gate determines how much of the previous hidden state should be passed on to the next time step. It maintains old memory while integrating new information, functioning similar to the combined forget and input gates in LSTM networks, but in a more streamlined form.

2.8.2 GRU Equations

The operations of a GRU at a time step t can be described using the following equations:

Reset Gate:
$$r_t = \sigma(W_r[h_{t-1}, x_t] + b_r). \tag{2.1}$$

Here, r_t is the reset gate vector, h_{t-1} represents the hidden state from the previous time step, x_t is the current input, and W_r and b_r are the weights and biases of the reset gate. The sigmoid function, σ, ensures that the gate outputs a value between 0 and 1.

Update Gate:
$$z_t = \sigma(W_z[h_{t-1}, x_t] + b_z). \tag{2.2}$$

The update gate vector, z_t, determines how much of the previous hidden state is passed on to the next time step.

Candidate Hidden State:
$$\tilde{h}_t = \tanh(W_h \cdot [r_t * h_{t-1}, x_t] + b_h) \tag{2.3}$$

The hidden candidate $\tilde{h}$ represents a potential new hidden state calculated using the reset gate and the new input.

Final Hidden state:
$$h_t = (1 - z_t)h_{t-1} + z_t \tilde{h}_t. \tag{2.4}$$

h_t, is a weighted combination of the previous hidden state h_{t-1} and the candidate hidden state $\tilde{h}_t$, controlled by the update gate.

2.8.3 How GRUs Work?

Reset Gate Functionality: The reset gate decides whether to combine the new input x_t with the previous hidden state h_{t-1} or to focus solely on the new input. A small reset gate value minimizes the influence of the previous hidden state, allowing the network to "reset" its memory when needed.

Update Gate Functionality: The update gate controls how much of the candidate hidden state $\tilde{h}_t$ should be combined with the previous hidden state h_{t-1} to form the new hidden state h_t. A high update gate value retains most of the information from the previous time step, enabling the network to remember relevant information over time.

2.8.4 Advantages of GRU Networks

- **Simplicity and Efficiency**: Compared to LSTMs, GRUs have fewer gates and parameters, resulting in reduced complexity and faster training times, making them ideal for scenarios with limited computational resources.
- **Comparable Performance**: For many tasks, GRUs achieve performance comparable to LSTMs, particularly in natural language processing (NLP) and speech recognition. GRUs are especially suitable for tasks with simpler data or where training time is a priority.
- **Faster Convergence**: GRUs have been found to converge faster than LSTMs due to their simpler architecture, making them a good choice when training time is limited.

2.8.5 Applications of GRU Networks

GRUs, like LSTMs, are employed in various sequence modeling applications:

- **Natural Language Processing (NLP)**:

  - *Text Generation*: GRUs can be used for text generation by predicting the next word in a sequence.
  - *Machine Translation*: Bidirectional GRUs (BiGRUs) model word dependencies in different languages for translation tasks.
  - *Speech Recognition*: GRUs process sequences of audio features for speech-to-text systems, converting spoken words into text.

- **Time Series Forecasting**: GRUs are effective in forecasting scenarios, such as stock price prediction or weather forecasting, where the goal is to predict future values based on prior data.
- **Anomaly Detection**: GRUs are used in time series anomaly detection, identifying patterns that deviate from normal sequences.

2.8.6 GRU Versus LSTM

- **Architecture**: GRUs have two gates (reset and update), whereas LSTMs have three gates (input, forget, and output). This makes GRUs simpler and potentially better suited for problems with smaller datasets or limited computational resources.
- **Performance**: Although both models perform similarly on many tasks, LSTMs may be more suitable for capturing long-term dependencies in complex sequences. GRUs, on the other hand, are preferred when training speed is important.

2.8.7 Limitations of GRU Networks

- **Limited Flexibility**: While GRUs are less complex and faster than LSTMs, their simpler architecture may limit their ability to capture complex sequential dependencies.
- **Empirical Decision**: Choosing between GRUs and LSTMs often involves trial and error. In some cases, GRUs may underperform compared to LSTMs, especially when fine-grained control over memory is needed.

2.9 Encoder–Decoder Networks

Encoder–decoder architecture is a deep learning architecture that has been proven to be very efficient when working with sequential data such as speech, text, and visual descriptions. This architecture is thus meant for mapping one sequence onto another and is thus appropriate where the input and the output are sequences of varying lengths (Fig. 2.5).

Fig. 2.5 Encoder–decoder architecture (created by the authors)

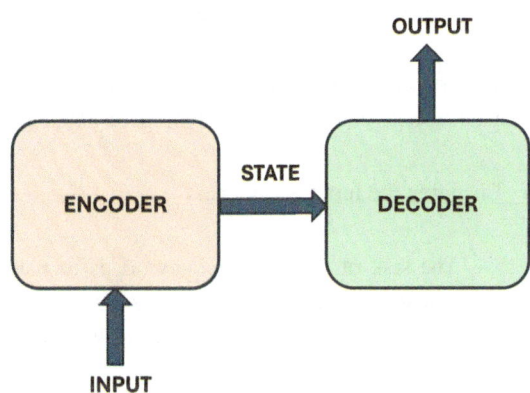

2.9.1 Components of the Encoder–Decoder Architecture

The encoder–decoder architecture consists of two main components:

- Encoder Network:

  - In the case of NLP, the encoder is used to transform the input sequence into a certain form. It consists normally of sequences of layers, these may include the Recurrent Neural Networks (RNNs), Long Short-Term Memory (LSTM) networks, or Gated Recurrent Units (GRUs).
  - The encoder processes an input sequence step by step, transforming it into a fixed-size vector, commonly referred to as a context vector, which encapsulates the information from the entire sequence.
  - The context vector aims to contain all the information that is required to produce the output sequence in question. It acts as an ending of the input sequence.

- Decoder Network:

  - The decoder network takes the encoder's context vector as its initial input and generates the output sequence.
  - Similar to the encoder, the decoder is often built using RNNs, LSTMs, or GRUs. It produces the output sequence incrementally, such as generating one word at a time in language translation tasks.
  - The decoder utilizes the context vector to begin generating the sequence and may optionally incorporate previous elements of the output sequence as inputs for predicting subsequent elements.

2.9.2 How the Encoder–Decoder Architecture Works?

To illustrate how this architecture works, let's take a real-life use case, which is actually translating a given sentence from a source language to another language.

- Encoding the Input Sequence:

  - The task of the encoder network is to read the input sentence in the source language word by word. As each word is received, it updates its hidden state to retain information about the sequence of words encountered so far.

2.9 Encoder–Decoder Networks

- The last fully hidden layer of the encoder receives the entire input sentence and becomes the context vector. Specifically, this vector summarizes the whole input sentence into a small vector where each word plays an important role.

- Generating the Output Sequence:

  - The process of translation in the target language starts with the transmission of the context vector to the decoder network.
  - The decoder produces the output step by step, with the help of the context vector and the decoder's hidden state to predict the target word sequence. For instance, it generates the first word of the translated sentence, then uses that word (along with the context) to predict the second word, and continues this process until the whole sentence is produced.

- Sequence-to-Sequence (Seq2Seq) Model:

  - The Seq2Seq model is specifically an encoder–decoder structure in which both encoder and decoder networks are recurrent neural networks or their variants, such as LSTMs and GRUs.
  - Applications of the Seq2Seq model include contexts that call for transforming an input sequence into an output sequence such as translating a line in one language to a line in another language in machine translation.

2.9.3 Key Challenges and Improvements

- Loss of Information in Context Vector:

  - The Context Vector losses the information because the entire input sequence is compressed into one fixed-size context vector and because of this, the decoder may provide quite incorrect outputs especially when the task is tough, such as in the translation of long sentences.

- Attention Mechanism:

  - Another problem with this approach: with vanishing gradients. Therefore, the attention mechanism is introduced. This mechanism allows the decoder to focus on different parts of the input sequence at each step of the output generation process, rather than relying solely on the context vector.

- With attention, the decoder can always gain access to the encoder's hidden states without much struggle hence be in a position to generate the right outputs for long and complex sequences.

2.9.4 Applications of the Encoder–Decoder Architecture

- Machine Translation:
  One common application is machine translation, where the encoder–decoder architecture translates a sentence from one language to another. The encoder processes the input sentence in the source language, and the decoder generates the corresponding translation in the target language.
- Speech Recognition:
  In speech recognition, the encoder takes an audio feature sequence as input and produces a context vector, while the decoder converts the context vector into a text sequence.
- Image Captioning:
  For image captioning, the encoder often uses a Convolutional Neural Network (CNN) to extract features from an image, which are then used as the context vector. The decoder, typically an RNN, generates a sequence of words to describe the image.

2.9.5 Advantages of the Encoder–Decoder Architecture

- Flexibility:
  The encoder–decoder architecture can handle a wide variety of sequence-to-sequence tasks, including translation, summarization, and more.
- Modularity:
  The encoder and decoder components can be designed independently, allowing the use of different types of networks (RNNs, LSTMs, GRUs, etc.) based on the task.

2.9.6 Limitations

- Difficulty with Long Sequences:
  The basic architecture struggles with long input sequences, as compressing all information into a single context vector can result in loss of information.

- Dependency on Sequential Processing:
  Both the encoder and decoder process sequentially, which can lead to inefficiencies in training and inference for long sequences.

2.9.7 Enhancements and Variations

- Attention Mechanism: As mentioned, the attention mechanism allows the decoder to access the entire sequence of encoder hidden states, improving the handling of long sequences.
- Transformers: Transformers are a newer architecture that builds on the encoder–decoder model but eliminates the reliance on RNNs. They use self-attention mechanisms to process sequences in parallel, improving efficiency and performance across many tasks.

2.10 Attention Mechanism

The attention mechanism is a major improvement over the simplest encoder–decoder solutions, especially for cases when the input sequence is extended, for example, in machine translation, text summarization, and many others. The motivation behind the attention mechanism is to make the model attend to specific positions within an input sequence during each step of decoding, while normally the input sequence is condensed into a fixed-dimensional vector. This approach minimizes the problem of information loss, which is especially so if one is dealing with long or complex input streams.

2.10.1 Traditional Encoder–Decoder Limitation

In the typical model of an encoder-decoder architecture: The encoder takes an input sequence (e.g., a sentence) and generates a fixed-size vector which is called the context vector or the thought vector. This vector is supposed to encode all the information from the input sequence through which the decoder decodes an output sequence. Although it has been turned into a fixed-length representation, in some cases for large sequences it falls short of coming up with a comprehensive summary of what it has taken in from the input sequence. This can result in poor output since the input often contains a lot of sub-topics that have to be catered for when generating the output.

2.10.2 Introduction of the Attention Mechanism

To overcome this limitation, the attention mechanism was designed into the RNNs. Introducing attention for the first time in the field of neural machine translation in their 2014 paper [7] "Neural Machine Translation by Jointly Learning to Align and Translate," it was demonstrated that attention completely alters how information is passed from the encoder to the decoder.

2.10.3 How the Attention Mechanism Works?

1. Dynamic Context Vector:

   a. Rather than relying on a fixed-size context vector as in the standard RNN decoding scheme, the attention mechanism provides the decoder with a new context vector at each decoding step. This context vector evolves based on the current state of the decoder and the corresponding part of the input sequence.

2. Attention Weights:

   a. Throughout the decoding process, the attention mechanism is able to produce a set of weights, which are referred to as attention weights. They express the level of impact each constituent of the input sequence has with regard to the progressed output step.
   b. For instance, if the decoder is at the time when it is generating the second word of the translated sentence, then the attention mechanism gives more focus on the part of the input sequence relating to this particular word in the translation.

3. Computing the Context Vector:

   a. The attention weights are then utilized to obtain a weighted average or sum of the encoder's hidden states. This weighted sum is the new context vector that the decoder uses to decode the next element of the output sequence.
   b. Through the dynamic calculation of the context vector at each iteration, the attentiveness brings out the circumstances whereby the decoder makes use of only the necessary portions of the input sequence to produce the next word or element of the output.

4. Alignment Between Input and Output:

   a. Another advantage of the attention mechanism is that it allows the model to learn an alignment between the input and output sequences. This means the model can identify which parts of the input sequence correspond to which parts of the output sequence, making its translations or outputs more accurate and context sensitive.

b. For instance, in machine translation, the attention mechanism can find out how various words in the source language are translated into the target language regardless of the word order between the two languages.

2.10.4 Visualization of Attention

The attention mechanism is depicted by attention maps, which are matrices that explain the alignment between the input and output sequences. Some uses of these maps include determining which part of the input the model is attending to, as it produces parts of the output. Figure 2.6: shows an example of an attention map demonstrating how one particular model has aligned elements of a French sentence (input on the top) with their corresponding English words (output at the bottom).

2.10.5 Applications and Impact

- **Machine Translation**: One of the biggest and most effective techniques, which has been utilized in the field of machine translation is that of the attention mechanism which helps the model in the process of translation to dedicate its attention to the most important words or phrases of the language being translated, rather than the entire text.
- **Text Summarization**: In the case of text summarization, the attention mechanism assists the model to pay attention to the important regions in the document, hence coming up with coherent and accurate summaries.
- **Image Captioning**: In image captioning, attention mechanisms can be utilized in order to select specific parts of an image when the model produces a caption. This enables the model to specify areas of the image more precisely in order to get the required picture.
- **Speech Recognition**: In speech recognition, we have encyclopedia structures that guide the model to the right position to look at while decoding the input feature and producing the textual transcription.

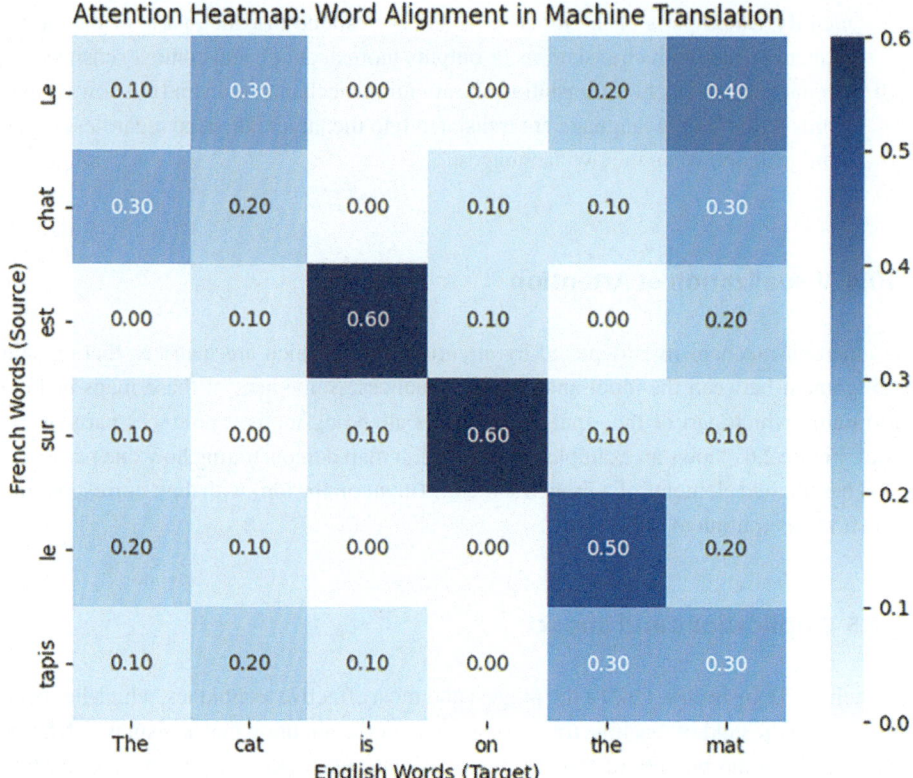

Fig. 2.6 Attention heatmap illustrating word alignment between a French sentence (source, listed along the y-axis) and its English translation (target, listed along the x-axis). Darker cells indicate higher attention values, showing how the model aligns words across languages despite differences in word order. This visualization demonstrates the ability of the attention mechanism to capture meaningful correspondences between source and target words in machine translation (created by the authors)

2.10.6 Variations and Extensions

- **Self-Attention**: This mechanism modifies what the model focuses on while processing the input sequence by attending to relationships within the same sequence, rather than between the encoder and decoder parts of the framework. This mechanism is especially important in transformer models, which have advanced the state of the art in NLP tasks.
- **Multi-Head Attention**: An extension of the attention mechanism that allows the model to focus on different parts of the input sequence simultaneously, capturing various aspects of the sequence while attending to all parts at once. This approach is commonly used in transformer models like BERT and GPT.

2.10.7 Advantages of the Attention Mechanism

- Better Handling of Long Sequences: This mechanism helps the model to cope with long input sequences that are common in various applications by permitting the decoder to learn and focus on a subset of the input sequence instead of operating on a fixed-size vector.
- Improved Performance: The application of the attention mechanism improves the performance of different models, especially in NLP and sequence-to-sequence models, by providing a more versatile and accurate representation of the input sequence.

2.10.8 Limitations

- Computational Complexity: Attention mechanism often resource-consuming, especially for longer input sequences as it requires attendance weight's computation for each sequence's element at each step of the decoding procedure.
- Interpretability: Although attention maps give some idea about what the model is attending to, it is not clear how these attention weights are derived, particularly in complex models.

2.11 Transformer Architecture

The Transformer architecture proposed by Vaswani et al. in their work [8] "Attention Is All You Need" in 2017, is a revolutionary step in the development of neural network architecture, especially when it comes to the manipulation of sequential data in NLP. Unlike the conventional seq2seq models that utilize either recurrent or convolutional modules, the Transformer architecture applies the self-attention modules enabling the processing of sequences on the input. Here's a detailed breakdown of its components and advantages:

2.11.1 Components of the Transformer Architecture

1. Encoder
   Function: For the entire input sequence, the encoder delivers several hidden representations in the encoding process.
   Structure: The encoder consists of multiple identical layers, each of which contains the following primary sub-layers:

- **Self-Attention Mechanism**: In this mechanism, the attention weights between each pair of input elements is calculated by the encoder so that the relevance of different parts of the input sequence can be compared.
- **Feedforward Neural Network**: Following the self-attention layer, the encoder applies a feedforward neural network to each position independently, transforming the self-attention output into more abstract states.
- **Normalization and Residual Connections**: Each sub-layer in the encoder is accompanied by a normalization operation and a residual connection to improve training capability and help in learning complex features.

2. Decoder
   Function: The decoder produces text from the encoded vector representations.
   Structure: Like the encoder, the decoder consists of several identical layers, but with an additional layer that reads the encoder's output:

- **Masked Self-Attention Mechanism**: This mechanism ensures that each position in the output sequence only attends to earlier positions, preventing future information from influencing current predictions.
- **Encoder–Decoder Attention**: This layer allows the decoder to attend to the encoder's output to produce a sequence that corresponds closely to the input sequence.
- **Feedforward Neural Network**: As in the encoder, a feedforward network processes the output of the attention layers.

2.12 Self-Attention Mechanism

The self-attention mechanism is one of the key innovations of the Transformer architecture. It allows the model to weigh the importance of different positions within the same input sequence, enabling it to learn long-range dependencies and contextual relationships without relying on recurrence.

- **Concept**: Self-attention enables each position in the input sequence to take into consideration every other position and then compute a weighted sum that capture contextual relationships, effectively enriching the representation at each position.

2.12 Self-Attention Mechanism

- **Process**:

  - **Attention Weight's Calculation**: In each position of the input, self-attention computes a set of weights that determine how much attention should be given to all other positions.
  - **Weighted Sum**: These weights are used to compute a weighted sum over all positions, and a measure reflecting contextual information from the entire sequence is obtained.

2.12.1 Advantages of the Transformer Architecture

1. **Parallel Processing**:

   - **Efficiency**: Unlike sequential models, such as recurrent models, the Transformer processes all positions in the sequence simultaneously. This parallel processing speeds up both training and inference, as computations for different positions can be performed concurrently.
   - **Scalability**: This parallelization is beneficial when training on large datasets and complex models.

2. **Enhanced Interpretability**:

   - **Attention Weight's Visualization**: The self-attention mechanism highlights the importance or relevance of each component of the input sequence for each position of the sequence output. These weights can be visualized as attention maps, providing insight into how the model's decisions are made and which parts of the input are most significant in generating each output.

3. **Improved Performance**:

   - **Handling Long-Range Dependencies**: Operating on the entire sequence at once makes it easier for the Transformer architecture to understand long-range dependencies and the relationships within the sequence. This capability is especially important for tasks such as Natural Language Processing (NLP), like machine translation, where capturing context over the entire sequence is needed for accurate translation.
   - **Flexibility**: The Transformer's ability to work with any sequence and its proven efficiency across multiple NLP applications make it highly flexible.

2.12.2 Transformer Architecture in Practice

1. **BERT (Bidirectional Encoder Representations from Transformers)** [8]: A model based on the Transformer encoder, aimed at understanding the context in the sequence of text by using information from both the beginning and the end. Focusing on problems like question-answering or text classification, BERT delivers the best performance.
2. **GPT (Generative Pre-trained Transformer)** [9]: A model based on the Transformer decoder architecture designed for generating contextually meaningful words and phrases. GPT is applied to tasks such as text generation, language modeling, and many more.
3. **T5 (Text-to-Text Transfer Transformer)** [10]: A single model that can be used for all NLP tasks, uniting them under the scope of text-to-text problems, implemented within the Transformer architecture.

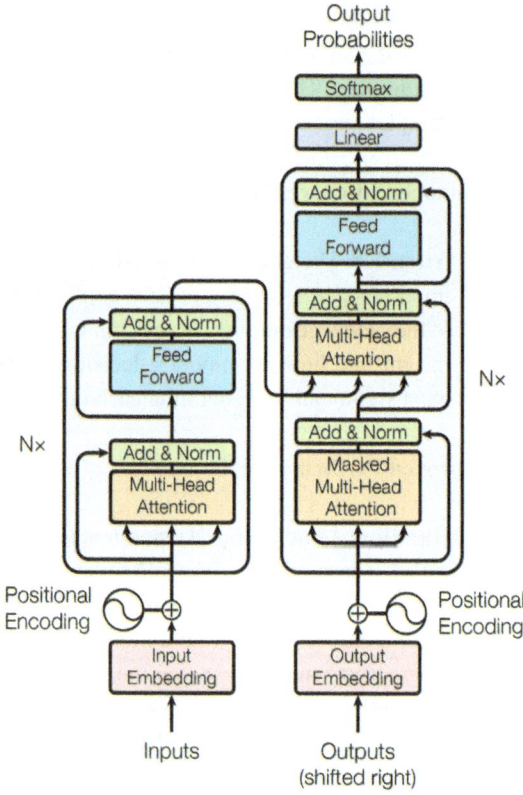

Fig. 2.7 Transformer architecture [8]

2.13 Large Language Models (LLMs)

Starting from 2018, most models of large language processing have employed what is known as the transformer structure, since this has improved the management of a variety of sequential information within deep learning. This marks a transition away from more conventional recurrent architectures like LSTM networks (Fig. 2.7).

2.13.1 Transformer Architecture Overview

The Transformer model, introduced in the research paper by Vaswani et al. in 2017 titled "Attention Is All You Need" [8], used self-attention techniques to extract and model sequences from data. In contrast, RNNs are sequential, reading through each item of the sequence one at a time, while Transformers take sequences in their entirety, making them efficient for large sequences.

2.13.2 General Architecture Properties

1. Tokenization

   - **Purpose**: Tokenization is the process of converting text into numeric values that the model can understand.
   - **Method**: Tokenizers convert text into lists of integers, which are methods of encoding text into tokens. These tokens can represent one character, one word, or one subword. For example, a tokenizer will represent "machine learning" as a sequence of two tokens: [machine, learning] and will assign a unique integer ID to each token like [1234, 5678].
   - **Training**: Pre-trained Tokenizers (PTs) are static and used to divide the corpus into blocks that the LLM is trained on. They are employed in text compression as well as encoding frequently used words or phrases into a single token.
   - **Token Size**: As seen in practice, tokenizers usually map one token into four or five characters or 0 to 3 bits. A typical English text corpus contains about 75 words per sentence.

2. Output

   - **Probability Distribution**: The output of an LLM is a probability distribution represented by the vocabulary of phrases or words that are common within the domain.
   - **Process**:

- **Vector Representation**: For each input sequence, the model returns a vector $y \in \mathbb{R}^V$ where V is the cardinality of the vocabulary.
- **Softmax Function**: Given the unnormalized logit vector x, the model applies the softmax function to produce the normalized probability distribution y. The softmax function transforms the logits into non-negative values whose sum equals one.
- **Predictions**: This is the probability distribution in the model for words contained in the vocabulary for the input text.

2.13.3 Examples of Recent LLMs

- GPT-4:

    - Overview: GPT-4—the fourth version of the generation model series produced by OpenAI.
    - Capabilities: Famous for its capability to produce text, which sounds natural like written by a human, answering questions, writing poetry, and even writing code. It builds upon the prior versions with improvements in both language generation and understanding.

- BERT:

    - Overview: BERT which stands for Bidirectional Encoder Representations from Transformers, BERT has been developed by Google.
    - Functionality: For instance, BERT take context from left and right directions in a text sequence where it is suitable for many NLP jobs such as question-answering task and sentiment analysis. Since BERT has the bidirectional model it is capable of analyzing the meaning of the word by looking at the context.

- T5:

    - Overview: Other developed by Google is T5 (text-to-text transfer transformer) which formats all the NLP tasks in text-to-text problems.
    - Capabilities: Strengths include, for example, translation, summarization, and question-answering since every task is considered as the text generation task. Using this format of computing, coming up with the model and applying it in various NL tasks becomes easier and unified.

- RoBERTa:

  - Overview: Another version of BERT that is further improved and optimized by Facebook.
  - Improvements: RoBERTa, an acronym for Robustly optimized BERT approach, is much better than BERT since it refines the model's training techniques and hyperparameters.

- Megatron:

  - Overview: Megatron has been created by NVIDIA and it aims to increase the size of the model while at the same time welcoming the efficiency of GPUs.
  - Purpose: Enables researchers to train models with billions of parameters which is virtually impossible on other architectures. Megatron is more of the line where more parameters are to be trained and more training data is proposed for a single model to take a higher position in terms of performance and ability to solve or enhance different NLP tasks.

2.14 Summary

The use of state-of-the-art neural networks has improved generative NPLs' ability to process natural language. To get around these problems, researchers have come up with probabilistic models like n-gram and hidden Markov models, and even more sophisticated models like neural network-based techniques.

Issues with scaling and vanishing gradient were encountered by earlier models. Some of these problems were found to be alleviated by RNNs, such as LSTM and GRU networks, which improved the network's capacity to handle sequential input and retain memory.

Some new techniques, mainly attention-based techniques such as the Transformer architecture, were introduced. The Transformer, due to self-attention mechanisms, has proven to be superior in a multitude of NLP tasks by efficiently attending to different segments of the input sequence significantly enhancing language modeling and other related applications.

Today, Large Language Models (LLMs) utilize these innovations to produce natural language text via the application of deep neural networks. Due to adopting ideas from RNNs, feedforward networks, and attention mechanisms, LLMs possess advanced text analysis and generation features, reflecting the development obtained in this rapidly developing field.

2.15 Multiple-choice Questions

In this section, you'll find a series of multiple-choice questions designed to test your understanding of key concepts in generative AI. Choose the correct answer for each question.

1. What does the n-gram model predict?

    (A) The next word based on the entire sentence
    (B) The likelihood of the next word based on the preceding $n - 1$ words
    (C) The probability of a word based on its frequency
    (D) The sentiment of the next word

2. Which model is commonly used to represent sequences of data with hidden states?

    (A) Hidden Markov Model (HMM)
    (B) Support Vector Machine (SVM)
    (C) k-Nearest Neighbors (k-NN)
    (D) Decision Tree

3. What is a major challenge faced by Recurrent Neural Networks (RNNs)?

    (A) Overfitting
    (B) Vanishing and exploding gradients
    (C) Lack of data
    (D) High computational cost

4. How do Recurrent Neural Networks (RNNs) handle sequential data?

    (A) By processing all data at once
    (B) By maintaining a state that summarizes previous inputs
    (C) By using fixed-size input windows
    (D) By storing all previous data

5. In an LSTM network, what is the purpose of the forget gate?

    (A) To determine how much new information is allowed into the memory cell
    (B) To control the retention of information in the memory cell
    (C) To manage the flow of information from the memory cell to the output
    (D) To update the weights of the network

2.15 Multiple-choice Questions

6. Which gate in an LSTM network determines how much new information is allowed into the memory cell?

   (A) Output gate
   (B) Input gate
   (C) Forget gate
   (D) Update gate

7. What distinguishes Gated Recurrent Units (GRUs) from Long Short-Term Memory (LSTM) networks?

   (A) GRUs use three gates, while LSTMs use two
   (B) GRUs use two gates, while LSTMs use three
   (C) GRUs are less effective in capturing long-term dependencies
   (D) GRUs cannot handle sequential data

8. What gates are used in Gated Recurrent Unit (GRU) networks?

   (A) Input and output gates
   (B) Reset and update gates
   (C) Forget and output gates
   (D) Input and forget gates

9. In the encoder–decoder architecture, what does the encoder network do?

   (A) Generates the output sequence from the context vector
   (B) Processes the input sequence to produce a fixed-length representation
   (C) Translates the input text into a different language
   (D) Adjusts the weights of the network during training

10. What role does the decoder network play in the encoder–decoder architecture?

    (A) It processes the input sequence
    (B) It generates the output sequence based on the context vector
    (C) It encodes the input data into a fixed-length vector
    (D) It handles input sequence errors

11. What problem does the attention mechanism address in traditional encoder–decoder models?

    (A) Difficulty in processing short sequences
    (B) Loss of information in long sequences

(C) Slow training times
(D) Inefficient use of computational resources

12. Who introduced the attention mechanism in neural networks?

    (A) Vaswani et al.
    (B) Bahdanau et al.
    (C) Hochreiter and Schmidhuber
    (D) Bengio et al.

13. What is a key feature of the attention mechanism in neural networks?

    (A) It allows parallel processing of data
    (B) It computes attention weights for all pairs of input elements
    (C) It reduces the size of the input sequence
    (D) It uses fixed-length vectors for all inputs

14. What advantage does the Transformer architecture offer over previous neural network designs?

    (A) Sequential processing of input sequences
    (B) Increased interpretability through visualizing attention weights
    (C) Simplified architecture with fewer layers
    (D) Reduced training time by limiting data processing

15. Which large language model is known for generating human-like text, answering questions, creating poetry, and writing code?

    (A) BERT
    (B) RoBERTa
    (C) Megatron
    (D) GPT-4

2.16 Answers

Below are the answers to the multiple-choice questions from the previous section:

1. (B) The likelihood of the next word based on the preceding $n - 1$ words
2. (A) Hidden Markov Model (HMM)

3. (B) Vanishing and exploding gradients
4. (B) By maintaining a state that summarizes previous inputs
5. (B) To control the retention of information in the memory cell
6. (B) Input gate
7. (B) GRUs use two gates, while LSTMs use three
8. (B) Reset and update gates
9. (B) Processes the input sequence to produce a fixed-length representation
10. (B) It generates the output sequence based on the context vector
11. (B) Loss of information in long sequences
12. (B) Bahdanau et al.
13. (B) It computes attention weights for all pairs of input elements
14. (B) Increased interpretability through visualizing attention weights
15. (D) GPT-4

References

1. Maryam Khanian Najafabadi. Sentiment analysis incorporating convolutional neural network into hidden markov model. *Computational Intelligence*, 40(2):e12633, 2024.
2. Yurun Wang, Yi Huang, Dongsheng Chen, Longyan Wang, Lingjian Ye, and Feifan Shen. Proxsta-lstm: A sparse representation for the attention-based lstm networks for industrial soft sensor development. *IEEE Access*, 2024.
3. Amjan Shaik, B Aruna Devi, R Baskaran, Satish Bojjawar, P Vidyullatha, and Prasanalakshmi Balaji. Recurrent neural network with emperor penguin-based salp swarm (rnn-eps2) algorithm for emoji based sentiment analysis. *Multimedia Tools and Applications*, 83(12):35097–35116, 2024.
4. S Hochreiter. Long short-term memory. *Neural Computation MIT-Press*, 1997.
5. Kyunghyun Cho, Bart Van Merriënboer, Caglar Gulcehre, Dzmitry Bahdanau, Fethi Bougares, Holger Schwenk, and Yoshua Bengio. Learning phrase representations using rnn encoder-decoder for statistical machine translation. *arXiv preprint* arXiv:1406.1078, 2014.
6. Junyoung Chung, Caglar Gulcehre, KyungHyun Cho, and Yoshua Bengio. Empirical evaluation of gated recurrent neural networks on sequence modeling. *arXiv preprint* arXiv:1412.3555, 2014.
7. Dzmitry Bahdanau. Neural machine translation by jointly learning to align and translate. *arXiv preprint* arXiv:1409.0473, 2014.
8. A Vaswani. Attention is all you need. *Advances in Neural Information Processing Systems*, 2017.
9. Jacob Devlin. Bert: Pre-training of deep bidirectional transformers for language understanding. *arXiv preprint* arXiv:1810.04805, 2018.
10. Colin Raffel, Noam Shazeer, Adam Roberts, Katherine Lee, Sharan Narang, Michael Matena, Yanqi Zhou, Wei Li, and Peter J Liu. Exploring the limits of transfer learning with a unified text-to-text transformer. *Journal of machine learning research*, 21(140):1–67, 2020.

LLMs and Transformers 3

By the end of this chapter, you will:

- **Unlock the Potential of Language Models**:
  Gain insight into how large language models, driven by advanced architectures, are transforming the way machines understand and generate human language.
- **Explore Why Transformers? The Evolution of Neural Architectures**
  Explore the reasons behind the development of the Transformer architecture, including its advantages over previous models like RNNs and LSTMs in handling sequential data.
- **Understand the Inside of a Transformer: A Structural Overview**
  Understand the components and design of the Transformer model, focusing on how it processes data through its encoder–decoder structure and attention mechanisms, and explore how these elements enable tasks such as translation and text generation.
- **Explore Decoding Attention: The Core of Transformers**
  Learn what self-attention means, why it is important to capture context relationships between tokens in a sequence, and how self-attention is capable of handling data independently in line with the structure of the Transformer.
- **Study Transformers in Perspective: Strengths and Shortcomings**
  Evaluate the possibilities that the Transformer has been offering, including its high scalability and parallelism for computations, as well as the existing weaknesses that were also observed, such as computational expense and the possibility of interpretational issues.

© The Author(s), under exclusive license to Springer Nature Switzerland AG 2026
D. Bhati et al., *A Beginner's Guide to Generative AI*, Synthesis Lectures
on Computer Science, https://doi.org/10.1007/978-3-031-84724-0_3

3.1 Unlock the Potential of Language Models

Since many applications of NLP rely on capturing the semantic meaning of language, language models constitute a key part of NLP [1]. Along with syntactic and semantic knowledge, these models employ grammar and production strategies to understand and generate contexts of the textual input, responding with a complexity, that closely resembles to human-like language generation.

The previous models suffered from the issues of handling long-distance dependencies and context understanding that limits them when dealing with complex sentences and coherent text passages. Two considerations emerged from these path dependencies: first and foremost, these path dependencies resulted in large language models (LLMs) that are much better than previous models because of their size, novel architecture, and improved performance. Some of the new features that LLMs offers enable them to process and generate text with a better emulation of human language. These include a better understanding of the context, tone, and intent during text processing.

As will be shown, LLMs rely on significant computational resource functionality and large datasets that allow them to understand the nuances of human language. During the training, different examples of data make them capable of generalizing their work and performing well on insignificant tasks such as summarizing lengthy documents, generating creative content, and handling a large number of questions. It also creates new opportunities within sectors such as healthcare, legal assistance, and customer service since it is highly important to understand the language accurately. To illustrate, let us begin by elucidating language models (LMs) through a less complicated example (Fig. 3.1).

Suppose that you are holding a conversation with one of your wise friends whose ability to understand your thoughts is impressive, even if you fail to complete your phrases. If you say, "I love eating...," your friend may fill in the blank with something like "pizza" or "ice cream" if that's what you two have been discussing. This is because your friend observes how you have previously interacted and can easily anticipate during text processing.

Similarly, language models also function in the same manner. With the help of certain archetypes, language models are also capable of creating and generating several other patterns as well. They are like spectators of ongoing actions and trained friends in computer science in order to understand and predict the language. They consider what you have typed so far and attempt to finish off the rest with the same feeling as when you try to guess the last word of a sentence your friend is forming in front of you.

For instance, if you were to type in "I want to purchase a new...," the model may complete the text as "phone," "car," or "book" depending on what the model has learned from a lot of text.

Due to their ability to identify and predict texts, the language models can be applied in numerous fields. They assist chatbots in answering your questions, writing assignments, translation between two languages, and giving summaries of articles. The aim is to ensure

3.1 Unlock the Potential of Language Models

Fig. 3.1 Example to demonstrate the predictive power of language models (LMs) (created by the authors)

that people find interactions they have with technologies as friendly and useful as they are when they are consulting a wise friend.

The language modeling took a significant leap forward with the advent of the Transformer architecture. The integrated paper "Attention Is All You Need" explains the recently developed attention mechanism that assesses the importance of one word concerned with all the others in a sequence. These two, along with other aspects of the architecture, have been responsible for the ability of the Transformers to rely on feedforward neural networks. Architecture to process the data in parallel, not sequentially has significantly enhanced the performance and its capacity to deal with a much more extensive dataset for training the model and the generation of the text more contextually relevant and accurate.

Moving forward, language models are anticipated to spearhead other developments in AI-based language comprehension and creation. It is expected that LLMs will help in upgrading human–computer interfaces, authoring, and robotics and come up with better solutions to arts and several creative fields such as writing, painting, and music. This is not limited to language translations; they can be applied to vision, robotics, and various AI investigations. Yet, with these capabilities lie ethical responsibilities, hence requiring proper ethics to be displayed. Training on different data we use might involve learning biases and passing them on to others, which is detrimental as it enhances stereotyping and the spreading of wrong information. Privacy is also an issue of these model sizes; the datasets used to train the models may contain private information. Solving them is quite vital to ensure that language models are used and developed in the right way, these challenges detecting biases, implementing

Fig. 3.2 Why transformers? (created by the authors)

methods to avoid or counter them, addressing and preventing potential misuse of these models and ensuring that the data privacy policies have been met.

3.2 Why Transformers? The Evolution of Neural Architectures

The Transformer Architecture is an improvement in NN models, particularly in capturing sequential information (Fig. 3.2). Before Transformers, Recurrent Neural Networks (RNNs) and their more evolutionarily developed counterpart, Long Short-Term Memory (LSTM) networks, used to be the go-to architectures for sequential data processing, like text or time series. RNNs and LSTMs also helped but Transformers brought certain changes that eliminated their drawbacks and added significant enhancements performance and effectiveness [2].

3.2.1 Challenges with RNNs and LSTMs

RNNs and LSTMs were developed to handle temporal dependencies in sequence data through the input sequential data one step at a time using a hidden state to preserve information through the sequence. Despite their advancements, these models had significant limitations:

- **Long-Range Dependencies**: Conventional RNNs in the past had a problem with long-term dependencies as they faced the vanishing and exploding gradients problem. These issues made it difficult for RNNs to memorize information over long sequences, leading to poor performance on tasks requiring distant context.
- **Sequential Processing**: RNNs work in a sequential manner which means that they cannot maximize parallel processing. This sequential approach can cause a lack of efficiency that was observed during the training and inference of the stages.
- **Complexity and Training Time**: RNNs came again with the problems of vanishing and exploding gradient and to overcome those issues LSTMs were proposed with additional cells for memory and control gates. However, LSTMs helped enhance the RNNs but had limitations when it came to working on very long sequences and the model was complex and time-consuming to train.

To overcome these challenges, transformers surfaced as new architecture for processing sequential data by incorporating superior design elements.

3.3 Inside the Transformer: A Structural Overview

The Transformer model that was proposed in the paper of Vaswani et al. [3] called "Attention Is All You Need" marks actual progress in the application of neural networks for sequential data processing [3]. Compared to the previous models like RNNs and LSTMs, the Transformer model does not use loops and relies on attention schemes and feedforward networks which are highly parallelizable.

3.3.1 Components of the Transformer Model

The Transformer architecture consists of two main components: the encoder and the decoder. These basic building blocks are essential for several tasks including machine translation and text generation are such. The design of the Transformer architecture effectively handles the inherent sequentiality of such tasks while preserving the context of the input sequences to generate accurate output. For more details on the specifics of the encoder–decoder system and its relevance to multiple NLP applications, read the description provided below.

3.3.2 Components of the Encoder–Decoder System

1. **Encoder**
   Although there are several variations of this design, the general function of the encoder is to transform the input sequence into a latent space that retains contextual information about the data. It consists of multiple encoder layers, each of which includes:

   - **Multi-Head Self-Attention**: Depending upon the input sequence, it builds up features-of-features which capture the relationships between words. The encoder is hence capable of comprehending several contextual relations due to the fact that each head in the multi-head attention mechanism attends to dissimilar features of the input.
   - **Feedforward Network**: This applies a non-linear transformation to each position in the sequence to enhance the model's capacity to learn complex patterns.
   - **Layer Normalization and Residual Connections**: Help stabilize and accelerate training by normalizing the outputs and allowing gradients to flow more easily through the network.

2. **Decoder**
   The decoder's task is to produce outputs from the encoded inputs in the form of the output sequence. It is structured similar to the encoder but includes additional components:

   - **Masked Multi-Head Self-Attention**: It helps to avoid watching one or several subsequent tokens in the decoder during the training stage and makes predictions based only on previous tokens.
   - **Multi-Head Attention Over Encoder Output**: Enables the decoder to concentrate on parts of the encoded input sequence that affect its predictions. This attention mechanism integrates the encoded information to produce an appropriate output compatible with the current context.
   - **Feedforward Network**: Takes the outputs of the attention mechanisms and converts them into the final predictions.
   - **Layer Normalization and Residual Connections**: These components aid the stability and efficiency of the training process, as observed in the encoder section (Fig. 3.3).

3.3 Inside the Transformer: A Structural Overview

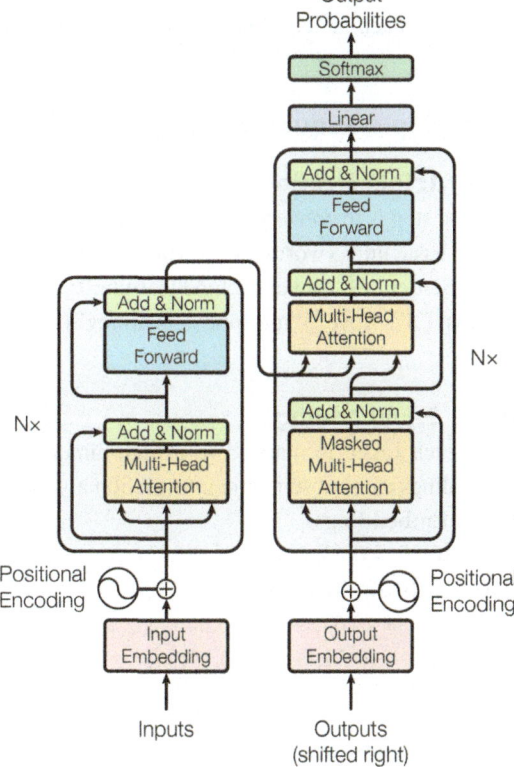

Fig. 3.3 The encoder–decoder structure of the Transformer architecture. Taken from "Attention Is All You Need" [3]

3.3.3 Detailed Process

1. Input Sequence Encoding
   The input sequence is first tokenized and converted into embeddings. These embeddings are then augmented with positional encodings to capture the order of the words. The encoded representations are processed through multiple encoder layers.
2. Output Sequence Generation
   The decoder takes the encoded representation and generates the output sequence one token at a time. The process includes:

   - **Masked Self-Attention**: The decoder uses masked attention to generate tokens in the output sequence. This ensures that each token is generated based only on previous tokens and not future ones.
   - **Encoder–Decoder Attention**: Integrates information from the encoder's output to generate contextually relevant tokens for the target sequence.
   - **Feedforward Network**: Refines the outputs to produce the final token predictions.

Let's break down the steps with a detailed example of translating the English sentence "I love machine learning" into French using a Transformer model.

3.3.4 Input Sequence Encoding

Step 1.1: Tokenization
The first step is to convert the input sentence "I love machine learning" into tokens. Tokens are the basic units (words or subwords) that the model processes.
Input Sentence: "I love machine learning"
Tokens: ["I", "love", "machine", "learning"]

Step 1.2: Embedding
Next, each token is mapped to a high-dimensional vector, called an embedding. These embeddings capture semantic information about the words.
Token Embeddings:
"I" $\to$ [0.25, −0.13, 0.48, ...]
"love" $\to$ [0.78, 0.01, −0.56, ...]
"machine" $\to$ [0.67, −0.22, 0.34, ...]
"learning" $\to$ [0.39, 0.55, −0.15, ...]

Step 1.3: Positional Encoding
Since Transformers don't have a built-in sense of order, positional encodings are added to the embeddings to introduce the order of words in the sentence.
Positional Encoding:
Position 1 (for "I") $\to$ [0.00, 0.98, 0.01, ...]
Position 2 (for "love") $\to$ [0.84, 0.54, 0.02, ...]
Position 3 (for "machine") $\to$ [0.91, 0.31, 0.03, ...]
Position 4 (for "learning") $\to$ [0.41, 0.73, 0.04, ...]
Combined Input Representation:
"I" $\to$ [0.25 + 0.00, −0.13 + 0.98, 0.48 + 0.01, ...] = [0.25, 0.85, 0.49, ...]
"love" $\to$ [0.78 + 0.84, 0.01 + 0.54, −0.56 + 0.02, ...] = [1.62, 0.55, −0.54, ...]
"machine" $\to$ [0.67 + 0.91, −0.22 + 0.31, 0.34 + 0.03, ...] = [1.58, 0.09, 0.37, ...]
"learning" $\to$ [0.39 + 0.41, 0.55 + 0.73, −0.15 + 0.04, ...] = [0.80, 1.28, −0.11, ...]

Step 1.4: Encoder Layers
These combined representations are then passed through the encoder layers, which consist of self-attention and feedforward networks. The encoder captures relationships between words, regardless of their positions in the sentence.

3.3 Inside the Transformer: A Structural Overview

Self-Attention: Calculates the importance of each word relative to others.
Example: The word "machine" might focus more on "learning" in the context of this sentence.
Feedforward Network: Processes the attention-weighted embeddings to produce a contextualized representation for each word.
Contextualized Representation:
"I" → [0.31, −0.05, 0.65, ...]
"love" → [0.85, 0.24, −0.32, ...]
"machine" → [0.71, −0.14, 0.40, ...]
"learning" → [0.50, 0.63, −0.10, ...]

3.3.5 Output Sequence Generation

Step 2.1: Decoder Input
The decoder starts generating the output sequence. Initially, it receives a special start token ⟨start⟩.
Decoder Input: ⟨start⟩

Step 2.2: Masked Self-Attention
The decoder predicts the next token by looking only at the previous tokens (initially just the ⟨start⟩ token). It cannot see future tokens, ensuring the sequence is generated in order.
Masked Self-Attention: Focuses only on previous tokens, not future ones.
After the first step: ⟨start⟩ → "J'"
After the second step: ⟨start⟩ J' → "aime"
After the third step: ⟨start⟩ J'aime → "l'apprentissage"
After the fourth step: ⟨start⟩ J'aime l'apprentissage → "automatique"

Step 2.3: Encoder–Decoder Attention
The decoder also attends to the encoder's output, ensuring the generated French tokens are contextually aligned with the original English sentence.
Encoder–Decoder Attention: Aligns tokens from the French sentence with the corresponding English words.
"I" → "J'"
"love" → "aime"
"machine learning" → "l'apprentissage automatique"

Step 2.4: Feedforward Network
The decoder's predictions are refined by a feedforward network, which ensures that the output tokens are accurate.

Step 2.5: Generated Output
In the last, the decoder generates the whole translated sentence token by token.

Generated Sentence: "J'aime l'apprentissage automatique"

For instance, a French-speaking user can create a post that states, "Je suis intéresser par l'intelligence artificielle" or translate to "I like machine learning."
This process continues until both the source message and a special end token ($\langle end \rangle$) are generated by the network.

Summary
Input: "I love machine learning"
Output: "J'aime l'apprentissage automatique"

It also shows how the Transformer model takes in the input sentence, encodes it to contextualized vectors, and then translates it into the relevant target language.

3.4 Decoding Attention: The Core of Transformers

The self-attention mechanism is at the core of the Transformer model and is used as a method of processing the input sentence so that the model can easily identify relationships between the words. While common architectures such as RNNs and LSTMs process input data sequentially, the Transformer gathers context and dependencies about words in a sequence from self-attention all at once, irrespective of their position of occurrence in the sequence.

3.4.1 Self-Attention Mechanism Explained

Self-attention, also known as scaled dot-product attention, helps the word in a given sentence look at other words and understand the context. This process is important for understanding the meaning of the sentence, as the significance of a particular word can change due to the presence of other words that come before or after it.

3.4 Decoding Attention: The Core of Transformers

Key Components of Self-Attention:

Queries, Keys, and Values:

Query (Q): This can be thought of as a question that the current word needs answered to determine its relationship with other words.

Key (K): This represents potential answers provided by all the other words in the sentence.

Value (V): This contains the actual information (or content) of the words that will be weighted based on the relevance determined by the attention mechanism.

Imagine you're trying to understand the word "bank" in the sentence "I went to the bank." The word "bank" (query) asks what context it's in—whether it's related to money or a river. The surrounding words like "went" and "to the" (keys) provide clues. Based on these clues, the meaning (value) related to "bank" is determined.

Attention Scores:
The model calculates how relevant each word (key) is to the word in focus (query) by taking the dot product of their vectors.

where the vectors $\mathbf{Q}$ and $\mathbf{K}$ have the same dimension and $\mathbf{K}^\top$ represents the transpose of $\mathbf{K}$.

$$\text{Attention Score} = \mathbf{Q} \cdot \mathbf{K}^\top,$$

These scores are then scaled by dividing by the square root of the dimension of the key vectors. This step is necessary to prevent the scores from becoming too large, which could destabilize the model during training.

$$\text{Scaled Attention Score} = \frac{\mathbf{Q} \cdot \mathbf{K}^\top}{\sqrt{d_k}}.$$

Analogy: If you're trying to evaluate how well each word in a sentence contributes to understanding "bank," you first compare each word using the (query-key dot product) and then scale the result so that the evaluation remains balanced.

Softmax Function:

The scaled scores are passed through a softmax function, which converts them into probabilities. These probabilities represent how much attention each word should pay to every other word.

$$\text{Attention Weights} = \text{Softmax}(\text{Scaled Attention Scores}).$$

This step ensures that the attention weights for all words add up to 1, making them easier to interpret.

Analogy: After comparing and scaling the importance of each word, softmax normalizes these scores into a set of probabilities that show how much attention each word should receive, ensuring that all probabilities add up to one, like splitting a pie among the words based on their importance.

Weighted Sum:

Finally, the attention weights are used to create a weighted sum of the value vectors. This sum produces a new, context-aware representation of the word in focus.
Let **V** represent the input vectors,

$$\text{Attention Output} = \sum(\text{Attention Weights} \times \mathbf{V}).$$

In essence, this output vector now carries information about the current word enriched by the most relevant parts of the other words in the sentence.

Analogy: After determining how much each word should influence the word "bank," you combine these influences (weighted sum) to form a nuanced understanding of what "bank" means in this specific context.

3.4.2 Example in Action

Consider the sentence: "The cat sat on the mat."

Query: When focusing on "sat," the query is "What is the action being taken?"

Key: The other words like "cat" and "mat" provide context—who is performing the action and where.

Value: The actual content (who and where) will be combined based on the relevance determined by the keys.

Output: The self-attention mechanism gives "sat" a context-aware meaning by considering the importance of "cat" and "mat."

3.4 Decoding Attention: The Core of Transformers

This process repeats for every word, allowing the Transformer to build a deep understanding of the entire sentence, regardless of the order or distance between words.

3.4.3 Significance of Self-Attention

1. **Capturing Contextual Relationships**: Self-attention allows the model to compute the importance of each word based on all the words in the sequence. This helps in identifying the kind of relationship and dependencies that exist between the words irrespective of the position of the word in a document or text.
2. **Parallel Processing**: Compared with RNNs, which process the sequence word by word, a Transformer deals with all the words at once. This parallel processing helps to reduce the training time and therefore increases the scalability of the system.
3. **Handling Long-Range Dependencies**: In self-attention, each word can access every other word in the sequence thus handling long-range dependencies which is a big problem with RNNs and LSTMs.
4. **Flexibility in Contextual Understanding**: The attention mechanism always learn from one context to another and helps the model to analyze data from another perspective.

Below is a detailed explanation of the self-attention mechanism with coding examples.

```python
import torch
import torch.nn as nn

class ScaledDotProductAttention(nn.Module):
    def __init__(self, d_model, dropout=0.1):
        super(ScaledDotProductAttention, self).__init__()
        self.d_model = d_model
        self.dropout = nn.Dropout(dropout)

    def forward(self, query, key, value, mask=None):
        # Calculate the attention scores
        scores = torch.matmul(query, key.transpose(-2, -1)) / (self.d_model ** 0.5)

        # Apply the mask (if provided) to ignore certain positions
        if mask is not None:
            scores = scores.masked_fill(mask == 0, float('-inf'))

        # Normalize the scores to get the attention weights
        attn_weights = nn.Softmax(dim=-1)(scores)
        attn_weights = self.dropout(attn_weights)

        # Compute the weighted sum of the values
        output = torch.matmul(attn_weights, value)
        return output, attn_weights
```

```python
class MultiHeadAttention(nn.Module):
    def __init__(self, embed_size, num_heads):
        super(MultiHeadAttention, self).__init__()
        self.num_heads = num_heads
        self.embed_size = embed_size
        self.head_dim = embed_size // num_heads

        assert (
            self.head_dim * num_heads == embed_size
        ), "Embedding size must be divisible by number of heads"

        self.values = nn.Linear(embed_size, embed_size)
        self.keys = nn.Linear(embed_size, embed_size)
        self.queries = nn.Linear(embed_size, embed_size)
        self.fc_out = nn.Linear(embed_size, embed_size)

        self.attention = ScaledDotProductAttention(embed_size)

    def forward(self, value, key, query, mask=None):
        N = query.shape[0]
        # Split the embedding into multiple heads
        value_len, key_len, query_len = value.shape[1], key.shape[1], query.shape[1]

        values = self.values(value).view(N, value_len, self.num_heads, self.head_dim)
        keys = self.keys(key).view(N, key_len, self.num_heads, self.head_dim)
        queries = self.queries(query).view(N, query_len, self.num_heads, self.head_dim)

        values = values.permute(0, 2, 1, 3)
        keys = keys.permute(0, 2, 1, 3)
        queries = queries.permute(0, 2, 1, 3)

        # Apply self-attention for each head
        out, attn_weights = self.attention(queries, keys, values, mask)
        out = out.permute(0, 2, 1, 3).contiguous().view(N, query_len, self.embed_size)

        return self.fc_out(out), attn_weights

# Example Usage
embed_size = 256
num_heads = 8
attention = MultiHeadAttention(embed_size, num_heads)

query = torch.randn(2, 10, embed_size)  # Batch size of 2, sequence length of 10
key = torch.randn(2, 10, embed_size)
value = torch.randn(2, 10, embed_size)

output, attn_weights = attention(value, key, query)
print(output.shape)  # Should print torch.Size([2, 10, 256])
print(attn_weights.shape)  # Should print torch.Size([2, 8, 10, 10])
```

Example 3.1 A multi-head self-attention layer

3.4.4 Code Explanation

Example 3.1 is the code for a multi-head self-attention layer, which is incorporated in all Transformer models and built using the PyTorch library. The ScaledDotProductAttention class computes the attention scores through scaled dot-product calculation [4]. The function that computes attention weights takes three inputs: query, key, and value tensors. It then performs matrix multiplication which is followed by scaling and softmax normalization. An optional mask can also be applied to exclude certain positions in the sequence. The attention weights are then utilized to calculate a weighted sum of the value tensor which characterizes the result of the attention mechanism.

The concept can be extended to multiple attention heads and that is where the MultiHeadAttention class comes into the picture. The proposed model begins by applying the elucidating linear transformations to the values, keys, and queries and employs a final linear layer for the output. In the forward pass, these embeddings are divided into several heads with each head is responsible for performing a different part of the attention mechanism. The resulting attention scores are normalized within head and then concatenated and passed through the final linear layer. Such a setup enables the model to capture different contextual relationships from different representation subspaces.

As stated in the example usage, the MultiHeadAttention instance is defined using the embedding size of 256 and 8 heads of attention. Structurally, it takes a random query, key, and value tensors with a batch size $= 2$ and a sequence length $= 10$. The output tensor which contains information of all the attention heads has the same sequence length as that of the input and the same embedding size as the word embeddings. The attention weight tensor contains the detailed weights of each of the attention heads for every word in the input sequence.

3.5 Transformers in Perspective: Strengths and Shortcomings

The concept of Transformer architecture has been game-changing in natural language processing (NLP) due to its good design and performance. Yet, just like any other tool, it has its major advantages and a great number of limitations. In the next two sub-sections, we discuss these factors in more detail at a micro-level.

3.5.1 Strengths of the Transformer Model

1. Scalability:

    - **Model Size and Training Data**: Transformers can scale to accommodate vast amounts of data and large model sizes. This scalability is crucial for leveraging

extensive datasets to improve performance. For instance, models like GPT-3 and GPT-4 have billions of parameters, which allows them to capture intricate patterns and nuances in language.
- **Performance Improvement**: As the model size and data scale up, the performance of Transformers generally improves, particularly in tasks requiring deep contextual understanding and generation. This scalability is a significant advantage in advancing the state of the art in NLP.

2. Parallel Processing:

- **Efficient Computation**: Unlike RNNs where processing is done in sequentially, the Transformer architecture can fully exploit parallel processing. As evidenced by its ability to process all the words in a sequence at once. This parallelism also improves both training and inference time.
- **Training Efficiency**: Also, due to the parallel nature of the Transformers, it is possible to use modern accelerators for neural network calculations, like GPUs and TPUs more effectively. This leads to less training time required as compared to other models like sequential models.

3. Attention Mechanism:

- **Contextual Understanding**: Therefore, in the Transformers, the attention mechanism allows the model to pay a particular attention to related parts of the input sequence while creating each part of the output. This enables the model to easily capture long-range dependencies and effectively establish contextual relationship.
- **Flexibility**: Attention mechanisms allow for flexible weighting of the relative contributions of different parts of the incoming sequence aiding both language understanding and generation.

4. Transfer Learning:

- **Pre-training and Fine-tuning**: This is especially true with transformers where problems in training deep learning models from scratch are well known. Instead models are first trained on huge datasets and then adjusted for a given task. This approach relies on the general language understanding gained during the pre-training process, enabling the model to learn and perform well on downstream tasks with comparatively less data.

3.5.2 Shortcomings of the Transformer Model

1. **Computational Demands**:

   - **Resource Intensive**: These models can be very costly making it difficult for small organizations or a single person to make use of these models.
   - **Inference Cost**: The depth of the attention mechanism and the number of parameters cause high cost at the estimation time (during inference), especially for models used in real-time applications.

2. **Training Time**:

   - **Extended Duration**: Training large Transformers can take a long time due to the volume of data and model complexity. This extended training time can postpone the introduction of new models and applications into the market.

3. **Interpretability Challenges**:

   - **Opaque Decision-Making**: The large number of parameters and layers in the models like Transformers makes them non-interpretable and hard to analyze. This opacity pertains to the decision-making process or the rationale behind certain outputs or results, making debugging and improving results challenging.
   - **Complexity of Attention Patterns**: While the attention mechanism offers enhanced contextual information, the distribution of attention weight may be intricate and not easily interpretable. This lack of transparency may make it difficult to verify model's behavior against expected outcomes.

4. **Bias and Fairness**:

   - **Inherent Biases**: Large, diverse datasets used to train the transformers may contain biased data, which the transformers would also mimic. This may result in solutions that reinforce prejudice or that are simply prejudiced, which creates ethical and social issues.
   - **Mitigation Efforts**: To overcome these biases, there is a need to conduct constant research and innovation in relation to tools to recognize, assess, and eliminate biases in model training and application.

3.6 Related Studies

Here are some papers and their contributions in the field:
1. **"Attention Is All You Need" by Vaswani et al.** [3]
   This pioneering paper proposed the Transformer model, replacing recurrence and convolution in sequence modeling. The Transformer architecture eliminates the use of recurrence, relying solely on self-attention to capture global correlations between input and output. This innovation allowed for parallelization during the training process, reducing the time needed to train models on large datasets. The Transformer architecture is the foundation of many high-performance models like BERT, GPT, and T5 and has achieved record-high accuracy in a range of NLP tasks, including translation, summarization, and question-answering.
2. **"BERT: Bidirectional Encoder Representations from Transformers" by Devlin et al.** [5]
   BERT introduced a novel approach to pre-training language models by processing text bidirectionally, from left to right and right to left, capturing contextual information from both directions. Through a pre-training technique called masked language modeling (MLM) and fine-tuning for specific tasks, BERT set new records in several NLP tasks. Its ability to capture the semantic meaning of words within sentences made it highly effective in tasks like sentiment analysis, named entity recognition, and question-answering.
3. **Brown et al. [6]. "GPT-3: Language Models are Few-Shot Learners"**
   GPT-3 represented a major advancement in language models with its 175 billion parameters, making it the largest of its kind at the time. This model demonstrated exceptional few-shot learning abilities, performing new tasks with minimal examples or without task-specific training. GPT-3's generalization capability was a paradigm shift in NLP, enabling it to excel in tasks like translation and summarization, and highlighting the power of large language models to imitate human language.
4. **"RoBERTa: A Robustly Optimized BERT Pretraining Approach" by Liu et al.** [7]
   RoBERTa built on the BERT model by improving the pre-training methodology. Key changes included fine-tuning on larger datasets, removing the NSP task, and training with larger mini-batches for longer periods. These improvements led to better performance than BERT across a range of tasks, showcasing that iterative enhancements can result in major gains in model accuracy. RoBERTa has been highly effective in NLP tasks like text classification, sequence labeling, and sentiment analysis.
5. **"T5: Exploring the Limits of Transfer Learning with a Unified Text-to-Text Transformer" by Raffel et al.** [8]
   T5 generalized all NLP tasks into a unified text-to-text format, allowing the model to be applied to any task framed in this manner. This approach enabled T5 to achieve state-of-the-art performance across many benchmarks, including translation, summarization, and question-answering. T5 relies on transfer learning, training on a massive text corpus

and fine-tuning for specific tasks, demonstrating that one model can perform well across a variety of tasks without task-specific architectures.

6. **"XLNet: Generalized Autoregressive Pretraining for Language Understanding" by Yang et al.** [9]

    XLNet introduced a generalized autoregressive pre-training approach that combined the strengths of autoregressive models like GPT and autoencoding models like BERT. It constructed bidirectional context using a permutation-based target, overcoming the limitations of masked language models. XLNet outperformed BERT on several NLP benchmarks, offering a more comprehensive understanding of language context and dependencies, while addressing issues such as word order and training on artificial mask tokens.

3.7 Summary

Altogether, the large language models (LLMs) based on the Transformer architecture can be referred to as a breakthrough in NLP. These models are good at comprehending and producing language by utilizing context well and by using big pre-training datasets. The performance in language tasks is more effective compared to other approaches and as such is the benchmark for natural language understanding and generality. Not only LLMs are proficient in generating syntactically well-formed and semantically relevant textual information but they are also capable of creating innovative data including narratives and dialogs. However, their fast adoption is a cause of ethical concerns, as they can maintain and strengthen prejudice, exacerbate fake news distribution, and other misuse impacts. Alleviating these concerns is critical in promoting the appropriate application of Artificial Intelligence as a technology. The future of LLMs is promising to become steady across several industries, including education, healthcare, and customer service to mention a few more involving an improved human–computer interface and language processing. With the future developments of the field, LLMs will be able to change the concepts of language and communication, thus produce a profound impact on all the spheres of human life.

3.8 Multiple-choice Questions

In this section, you'll find a series of multiple-choice questions designed to test your understanding of key concepts in generative AI. Choose the correct answer for each question.

1. Which aspect of the Transformer architecture allows it to avoid the vanishing gradient problem commonly encountered in RNNs?

    (A) Recurrent connections
    (B) Multi-head self-attention
    (C) Feed-forward networks
    (D) Layer normalization

2. What is the primary benefit of using positional encodings in the Transformer model?

    (A) To enhance the parallel processing of sequences
    (B) To capture the order of words in a sequence
    (C) To reduce the computational complexity of attention mechanisms
    (D) To improve the interpretability of attention patterns

3. How does the use of multi-head attention in Transformers improve model performance?

    (A) By allowing the model to process multiple sequences simultaneously
    (B) By enabling the model to focus on different parts of the sequence with separate attention heads
    (C) By reducing the number of parameters in the model
    (D) By simplifying the computation of attention weights

4. In the context of Transformer models, what does the term "masked attention" refer to?

    (A) Attending to all tokens in the sequence without restriction
    (B) Preventing the model from attending to future tokens during training
    (C) Applying attention only to tokens in the middle of the sequence
    (D) Masking the input embeddings with noise

5. Why does the Transformer model use a feed-forward network in addition to self-attention layers in each encoder and decoder block?

    (A) To manage long-range dependencies more effectively
    (B) To perform non-linear transformations and enhance feature representation
    (C) To simplify the training process by reducing computational demands
    (D) To maintain sequential order information in the data

6. Which component of the Transformer architecture helps in handling the large number of parameters and stabilizing the training process?

    (A) Multi-head attention
    (B) Positional encoding
    (C) Residual connections
    (D) Convolutional layers

3.8 Multiple-choice Questions

7. How does the attention mechanism in Transformers handle variable-length input sequences?

    (A) By padding sequences to a fixed length
    (B) By applying attention weights to all positions in the sequence
    (C) By dynamically adjusting the sequence length during training
    (D) By using convolutional layers to process sequences

8. What challenges associated with traditional sequence models does the Transformer's parallel processing address?

    (A) The difficulty in capturing long-range dependencies
    (B) The inefficiency in handling variable-length sequences
    (C) The slow training and inference times due to sequential processing
    (D) The inability to process large datasets effectively

9. In the self-attention mechanism, how is the relevance of different tokens in a sequence determined?

    (A) By computing the dot product between token embeddings
    (B) By scaling the dot product of query and key vectors
    (C) By averaging the token embeddings
    (D) By applying a convolutional filter to the sequence

10. Which aspect of the Transformer's encoder-decoder architecture enhances its ability to perform tasks like machine translation?

    (A) The use of recurrent connections for long-range context
    (B) The combination of masked attention and encoder-decoder attention
    (C) The application of pooling layers to summarize sequences
    (D) The use of unsupervised learning during training

11. What role do residual connections play in the Transformer's architecture?

    (A) They facilitate the computation of attention weights
    (B) They add noise to the training data to prevent overfitting
    (C) They help gradients flow through the network and stabilize training
    (D) They adjust the dimensionality of the input embeddings

12. How does the Transformer's use of feed-forward networks in each layer contribute to its functionality?

    (A) By providing a mechanism for sequence alignment
    (B) By performing non-linear transformations on the attention outputs
    (C) By replacing the need for attention mechanisms
    (D) By normalizing the token embeddings

13. In what way does the Transformer architecture's design contribute to its scalability compared to earlier models?

    (A) By reducing the number of layers required for training
    (B) By enabling parallel processing of all tokens in a sequence
    (C) By using recurrent layers to maintain state across tokens
    (D) By simplifying the attention mechanism to only focus on adjacent tokens

14. What specific advantage does the use of scaled dot-product attention provide in the Transformer's self-attention mechanism?

    (A) It reduces the dimensionality of the attention outputs
    (B) It stabilizes gradients during training by scaling the attention scores
    (C) It improves the interpretability of the attention weights
    (D) It enhances the model's ability to handle variable-length sequences

15. Which feature of Transformers is primarily responsible for their ability to handle long-range dependencies more effectively than RNNs?

    (A) The use of convolutional layers
    (B) The self-attention mechanism that allows direct access to all tokens
    (C) The incorporation of memory cells in the network
    (D) The use of fixed-size attention windows

3.9 Answers

Below are the answers to the multiple-choice questions from the previous section:

1. (B) Multi-head self-attention
2. (B) To capture the order of words in a sequence
3. (B) By enabling the model to focus on different parts of the sequence with separate attention heads
4. (B) Preventing the model from attending to future tokens during training
5. (B) To perform non-linear transformations and enhance feature representation
6. (C) Residual connections
7. (B) By applying attention weights to all positions in the sequence
8. (C) The slow training and inference times due to sequential processing
9. (B) By scaling the dot product of query and key vectors
10. (B) The combination of masked attention and encoder–decoder attention
11. (C) They help gradients flow through the network and stabilize training
12. (B) By performing non-linear transformations on the attention outputs
13. (B) By enabling parallel processing of all tokens in a sequence

14. (B) It stabilizes gradients during training by scaling the attention scores
15. (B) The self-attention mechanism that allows direct access to all tokens

References

1. Valentina Alto. *Building LLM Powered Applications: Create intelligent apps and agents with large language models.* ISBN-13:978-1835462317, ISBN-10:1835462316. Packt Publishing Ltd, 2024.
2. Hyunwook Park, Yifan Ding, Ling Zhang, Natalia Bondarenko, Hanqin Ye, Brice Achkir, and Chulsoon Hwang. High-speed channel transformer: A scalable transformer network-based signal integrity simulator. *IEEE Transactions on Electromagnetic Compatibility*, 2024.
3. A Vaswani. Attention is all you need. *Advances in Neural Information Processing Systems*, 2017.
4. Yongping Du, Bingbing Pei, Xiaozheng Zhao, and Junzhong Ji. Deep scaled dot-product attention based domain adaptation model for biomedical question answering. *Methods*, 173:69–74, 2020.
5. J. Devlin, M. W. Chang, K. Lee, and K. Toutanova. Bert: Pre-training of deep bidirectional transformers for language understanding. In *Proceedings of the 2019 Conference of the North American Chapter of the Association for Computational Linguistics: Human Language Technologies*, volume 1, pages 4171–4186, 2019.
6. Tom B Brown. Language models are few-shot learners. *arXiv preprint* arXiv:2005.14165, 2020.
7. Y. Liu, M. Ott, N. Goyal, J. Du, M. Joshi, D. Chen, O. Levy, M. Lewis, L. Zettlemoyer, and V. Stoyanov. Roberta: A robustly optimized bert pretraining approach. *arXiv preprint* arXiv:1907.11692, 2019.
8. Colin Raffel, Noam Shazeer, Adam Roberts, Katherine Lee, Sharan Narang, Michael Matena, Yanqi Zhou, Wei Li, and Peter J Liu. Exploring the limits of transfer learning with a unified text-to-text transformer. *Journal of machine learning research*, 21(140):1–67, 2020.
9. Z. Yang, Z. Dai, Y. Yang, J. Carbonell, R. Salakhutdinov, and Q. V. Le. Xlnet: Generalized autoregressive pretraining for language understanding. In *Advances in Neural Information Processing Systems*, volume 32, pages 5753–5763, 2019.

The ChatGPT Architecture: An In-Depth Exploration of OpenAIs

4

By the end of this chapter, you will:

- **Explore Conversational Language Models**: What are conversational language models, such as ChatGPT, and how do they redefine the approach of human–computer conversational dialogs?
- **Learn about the Evolution of GPT Models**: We describes the evolution of GPT models from the time they were first introduced to the present day, focusing on the evolution of the models' architecture and how it has improved their ability to create realistic text.
- **Understand Architecture of ChatGPT**: Go deeper into the architecture of ChatGPT, reviewing what parts it consists of and which design decisions make it capable of effectively processing and creating conversational responses.
- **Formulate Pre-training and Fine-Tuning in ChatGPT**:

    - Pre-training: Understand the context of pre-training, where the model undergoes through training with huge text data in order to get the fundamental knowledge about the language.
    - Fine-Tuning: Focus on Certain Roles: Learn about how fine-tuning helps to enhance the performance of ChatGPT on specific tasks or in certain contexts to improve the efficiency of a given application or domain and its effectiveness in providing proper and appropriate answers.
    - Continuous Learning and Iterative Improvement: Learn how improvements and continuous learning add up to the development of the model as well as the ways in which this model manages to address new problems.

- **Contextualize Embeddings in ChatGPT**: Learn how ChatGPT uses contextual embeddings and how the technology allows the model to give context-based answers.
- **Navigate Response Generation in ChatGPT**: Find out how ChatGPT arrives at its responses and how it transforms an input into generating a text that follows the conversational tone appropriately.

    – Handling Biases and Ethical Considerations: Addressing Biases in Language Models: Look at the issues related to the biases in the language models and their effects on the concepts of bias and fairness.
    – OpenAI's Efforts to Mitigate Biases: Analyze the measures taken by OpenAI to prevent biases, as well as various initiatives adopted by it for the proper and ethical use of AI.

- **Understand Strengths and Limitations**: Assess the advantages of ChatGPT-based on its ability to engage in conversations and its versatility, and look into the weaknesses of this tool, including the possibility of it being prejudiced and the challenges of ethical application of this artificial intelligence product.

4.1 Exploring Conversational Language Models

The past few years have been revolutionary for Natural Language Processing (NLP), ushering in a new era of conversational agents [1]. These deep learning and big data-driven models have evolved from simple artificial intelligence programs that merely mimic conversations into sophisticated systems capable of engaging in complex dialogs with people. Among these advancements, ChatGPT stands out as a remarkable example of the extraordinary capabilities achieved with cutting-edge language models. Released by OpenAI, ChatGPT

4.1 Exploring Conversational Language Models

is built on the Generative Pre-trained Transformer (GPT) architecture, representing the next generation of conversational AI solutions.

Earlier conversational agent models, which relied on rule-based formalisms or relatively simple neural networks, faced significant challenges in maintaining coherence across consecutive conversations. As a result, interactions often lacked connection and relevance to one another. In contrast, the transformer-based framework employed by ChatGPT represents a substantial advancement, enabling more sophisticated and context-aware dialog interactions compared to models used in earlier experiments, such as Microsoft's conversational AI Agents.

Studies have shown that earlier models frequently struggled to sustain fluency, relevance, and logical coherence across interactions. ChatGPT, however, excels in these areas, consistently maintaining high-quality dialog. This improvement is largely attributed to the self-attention mechanism within its architecture, which allows the model to selectively focus on specific parts of the input text based on the context and intent of the interaction.

ChatGPT has been applied across various domains, including customer service, where it effectively responds to diverse questions, and education, where it assists in tutoring by providing explanations and answering a wide range of queries. This versatility stems from its pre-training phase, during which it was trained on extensive volumes of text data. As a result, ChatGPT has developed a broad understanding of language, enabling it to perform these tasks fluently and with high relevance.

Let's learn about ChatGPT with a simple example: Imagine the following customer support scenario. A user is trying to think of a reason why he or she is losing Internet connection. A conventional chatbot may lack continuity where it is used and often fail to understand which conversation is which; thus, further questions may be repeated. ChatGPT, on the other hand, can recall details of the user's problem in successive sessions. For example, after the user explains that their issue is that "My internet connection drops every few hours," ChatGPT can propose that the user examines the router settings, and then in the further dialog ask "Have you located the reset button on the back of your router?" If it does not, ChatGPT may then lead the user through further steps toward problem-solving, while maintaining the context of the conversation and previous methods tried. This is something that the older models have always lacked, the ability to remember the context and give the most relevant and coherent answer.

There are various innovations applied within the architecture of ChatGPT [2]. The attention and feedforward networks are the core of the transformer architecture employed by it, which help in processing and generating language. It allows the model to rank the contribution of words in a sentence and, as such, enhance its capacity to learn and exhibit context and coherence. Furthermore, the model is trained with a large number of datasets, which allows the model to have general knowledge on various topics and, at the same time, enables it to provide specific responses about various fields. Nevertheless, one cannot disregard the weak points of ChatGPT as a tool that is designed to perform a variety of functions. Despite the fairly proficient performance of the employed model, there are certain restrictions related to its applicability, such as fitting into incorrect or ambiguous contexts and certain biases in the data that were used to train the model. For example, ethically, there is an issue with the idea that the model may produce prejudice or offensive content if it is not adequately controlled. Furthermore, such models' application underlines a vast number of ethical problems, like the risk of spreading false information, violation of privacy rights, and the possibility of using the models for spreading hateful content. Mitigating these challenges is vital to the proper application of conversational AI applications.

In this chapter, we give readers a deeper insight into ChatGPT with specific details on the system's architecture and the top factors responsible for its performance. Understanding such elements, it is possible to explain how the modern language models are shaping the further evolution of human–computer interaction, thus providing the foundation for building more sophisticated conversational interfaces. We shall also look at the consequences of these technologies and the impacts that these technologies will pose on various fields, as well as the ethical issues that need to be met to make good use of these technologies.

4.2 The Evolution of GPT Models

The recently released Generative Pre-Trained Transformer (GPT) models have revolutionized natural language processing (NLP) and artificial intelligence (AI) as a whole. The evolution of GPT models developed by OpenAI highlights significant advancements in AI research, with each successive version introducing improvements that enhanced the models' ability to generate realistic and coherent text. From the initial version to the present day, the progression of GPT models demonstrates not only architectural enhancements but also record-breaking performance across a wide range of NLP tasks.

4.2.1 GPT-1: The Foundation

Released in 2018, GPT-1 marked the beginning of a new era in NLP. It introduced the use of the Transformer architecture, as presented in 2017 paper "Attention is All You Need" [3] of Vaswani et al. Unlike earlier models that relied on RNNs or CNNs, GPT-1 utilized a unidirectional Transformer model. This architecture, particularly the self-attention mechanism, enabled the model to consider the entire context of a sentence while generating words, resulting in more contextually relevant text generation (Fig. 4.1).

Key features of GPT-1:

- **Transformer Architecture**: GPT-1 employed self-attention to capture contextual dependencies between the words in a sentence. This approach addressed limitations in Recurrent

4.2 The Evolution of GPT Models

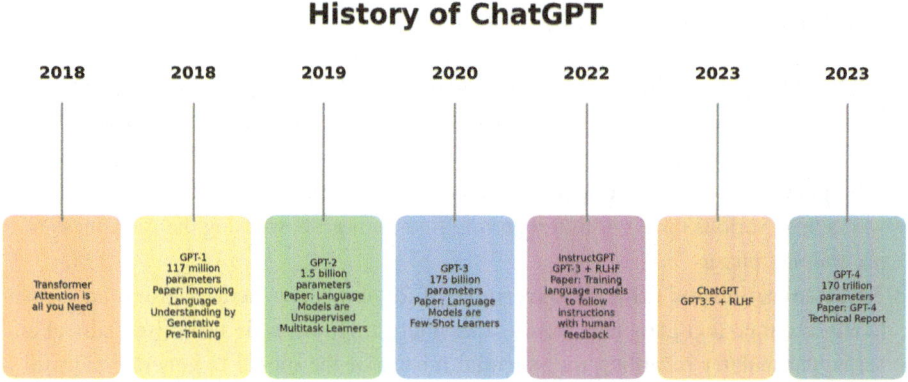

Fig. 4.1 History of GPT models (created by the authors)

Neural Networks (RNNs) and Convolutional Neural Networks (CNNs), which struggled with long-range or distant dependencies.

- **Pre-training and Fine-tuning**: A key feature of GPT-1 was its two-stage training process: pre-training on a large corpus of text followed by fine-tuning for specific tasks. This enabled the model to generalize across various NLP problems, effectively tackling classification tasks while providing a foundation for experimentation and reinforcement learning analyses.
- **Unsupervised Learning**: One of the most significant advancements was the model's ability to learn from unannotated text, as its unsupervised learning approach demonstrated exceptional potential for NLP applications.

Impact and Limitations:

While GPT-1 was the first model to showcase the potential of large-scale pre-training, its unidirectional approach limited its performance in tasks requiring a deeper understanding of input context. Additionally, the model's relatively small size, with only 117 million parameters, constrained its ability to capture the more intricate aspects of language.

4.2.2 GPT-2: Scaling Up

In 2019, OpenAI released GPT-2, a new generation of the GPT model that was significantly larger and more powerful than its predecessor. With 1.5 billion parameters, GPT-2 outperformed GPT-1 by generating more fluent and contextually relevant text from the given input. Trained on a massive dataset of 40 GB of Internet text, GPT-2 represented a major advancement in NLP, showcasing human-like text generation across various domains.

Key Features of GPT-2:

- **Larger Model Size**: GPT-2 incorporated significantly more parameters, enabling it to capture intricate patterns in data. This led to the generation of text that was more coherent, logical, and contextually sound.
- **Unsupervised Learning at Scale**: The model demonstrated its versatility by performing well across various tasks without requiring fine-tuning, showcasing the robustness of its pre-training phase.
- **Text Generation and Ethical Concerns**: GPT-2's ability to generate sensible and semantically accurate text garnered considerable interest. However, it also raised ethical concerns, particularly regarding the potential misuse of the model to generate harmful or misleading content, such as fake news or malicious materials.

Impact and Limitations:
The release of GPT-2 sparked widespread discussions about the applications of Artificial Intelligence. Its remarkable ability to generate highly realistic text raised concerns about its potential misuse, such as spreading fake news, creating deep fakes, and other malicious activities. Initially, OpenAI withheld the full model to assess these risks, marking a pivotal moment in the debate over the responsible development and deployment of such technologies.

4.2.3 GPT-3: The Leap Forward

Launched in 2020, GPT-3 is the largest and most complex GPT model to date, with a total of 175 billion parameters. GPT-3 marked a significant leap forward in the prominence of language models, demonstrating that they could perform a wide range of tasks with little to no need for additional fine-tuning or task-specific training. This ability to effectively handle various tasks made GPT-3 a landmark achievement in the development of AI.

Key Features of GPT-3:

- **Few-Shot Learning**: GPT-3 demonstrated the ability to perform tasks with minimal exposure, effectively generalizing from a small number of examples. This was in contrast to earlier models, which required large amounts of task-specific data.
- **Increased Scale**: With its enormous size, boasting 175 billion parameters, GPT-3 could store vast amounts of information, enabling it to perform well across a wide range of domains and tasks.
- **Zero-Shot and One-Shot Learning**: GPT-3 was capable of performing tasks it had never been explicitly trained on, requiring only a clear set of instructions in natural language. Its versatility and stability were evident, as it effectively generalized from minimal input.

- **API and Accessibility**: By releasing GPT-3 as an API, OpenAI allowed developers to integrate its functionality into various products, broadening the impact and accessibility of this powerful model.

Challenges and Considerations:

Nonetheless, several issues arose due to GPT-3's large size. For example, the model sometimes struggled to complete certain input sentences. The computational power required for training and running these models was substantial, raising concerns about the sustainability and accessibility of such large models. Additionally, despite the progress, GPT-3 still faced challenges with bias and factual inaccuracies, highlighting the continued importance of studying model safety and fairness.

4.2.4 GPT-4 and Beyond: Pushing Boundaries

Since 2024, the advancements in GPT-4 and subsequent iterations of language models have redefined the potential capabilities of these models. These newer models have incorporated enhanced techniques, including multimodal learning, where textual data is combined with other forms of data, such as images and sounds. This enables the models to engage in much richer and more complex interactions. The ongoing updates to GPT models reflect both the growing technical development and the increasing focus on ethical considerations.

Key features of GPT-4 and beyond:

- **Multimodal Capabilities**: GPT-4 introduced the concept of multimodal embeddings, incorporating data such as images, audio, and even video, to enhance the model's interpretability and generative abilities. This multimodal approach enables more sophisticated interactions in terms of context, expanding the range of potential applications for the model.
- **Enhanced Fine-tuning Techniques**: New fine-tuning methods have been developed to refine the model's outputs. Techniques such as reinforcement learning from human feedback (RLHF) and domain-specific fine-tuning help minimize the risk of generating toxic or biased content.
- **Ethical Considerations and Bias Mitigation**: As GPT models' ability to generate and disseminate content grows, ethical concerns have become more prominent. GPT-4 includes improved mechanisms for bias detection and reduction, enhanced content filtering, and added safety measures to ensure the generation of socially responsible content.
- **Domain Adaptation and Specialization**: The latest GPT models exhibit improved generalization across various domains compared to earlier versions. They can be fine-tuned for specific sectors such as healthcare, finance, and law, where domain knowledge and precision are critical.

- **Long-Form and Interactive Capabilities**: Recent versions of GPT have improved the ability to generate longer, coherent responses and better handle conversational context, enabling comprehension across a sequence of messages in an ongoing dialog.
- **Efficient Training and Deployment**: Advancements in training, including the use of sparse attention mechanisms and model distillation, have enabled the deployment of larger models in resource-constrained environments without compromising performance.

4.3 Architecture of ChatGPT

ChatGPT is a new-generation conversational model based on GPT (Generative Pre-trained Transformer) architecture which is developed to produce fluent and contextually appropriate responses. Some of its components are built from the Transformer model, but presidents are made to conform to this model with added extra provisions to facilitate conversational responses generation. This section discusses the basic components and the decisions that were made in architecture that allow the model behind ChatGPT to perform well in dialog, and show how these architectural features help in natural language understanding and generation. To refresh our memory about the Transformer model which is described in this chapter in more detail, let's consider its main points.

4.3.1 Transformer Architecture: The Backbone

The Transformer architecture, which forms the foundation of ChatGPT, processes entire sequences of text at once rather than token by token. This approach allows it to capture all the relationships that may exist within the text, providing a more comprehensive understanding of the input.

- **Self-Attention Mechanism**: The key feature of the Transformer architecture is the self-attention mechanism, which allows the model to assign specific weights to each word in a sequence, indicating its importance in relation to other words. This mechanism enables the model to capture long-range dependencies, which are essential for establishing context and context-awareness when responding in conversation. The self-attention mechanism calculates the attention between every pair of words in a sentence, allowing the model to focus on relevant input fragments at specific points in time.
- **Multi-Head Attention**: To enhance the model's ability to focus on multiple aspects of the input text simultaneously, the Transformer uses multi-head attention. In ChatGPT, multiple attention heads work in parallel, with each head processing a different part of the input sequence independently. This parallelism enables the model to link words and phrases in a more complex and nuanced way, improving its ability to understand and

generate better responses. Each attention head processes the input differently, allowing multiple perspectives on the same sequence before combining the results.
- **Feedforward Neural Networks**: To further improve the model's capability to process the input text, the Transformer includes feedforward neural networks in each layer. These networks process the output from the multi-head attention mechanism, enabling the model to transform the sequence representations and capture complex relationships in the data.
- **Layer Normalization and Residual Connections**: To stabilize training and enhance generalization, each layer in the Transformer architecture incorporates layer normalization and residual connections. Layer normalization normalizes the input to each sub-layer, ensuring consistent activations and reducing the risk of overfitting. Residual connections allow the model to retain information from previous layers by adding the input to the output, helping prevent the vanishing gradient problem and ensuring effective backpropagation in deep networks. These components are crucial for maintaining stability and performance as the network depth increases (Figs. 4.2 and 4.3).

4.3.2 Architecture of ChatGPT: Decoder-Only Structure

- **Unidirectional Attention**: In ChatGPT's decoder, an individual token (word) can only take or observe tokens that are preceding it in the sequence. That means, for instance, if the model is at the stage of "I love ice cream" and is at "ice," it only has information

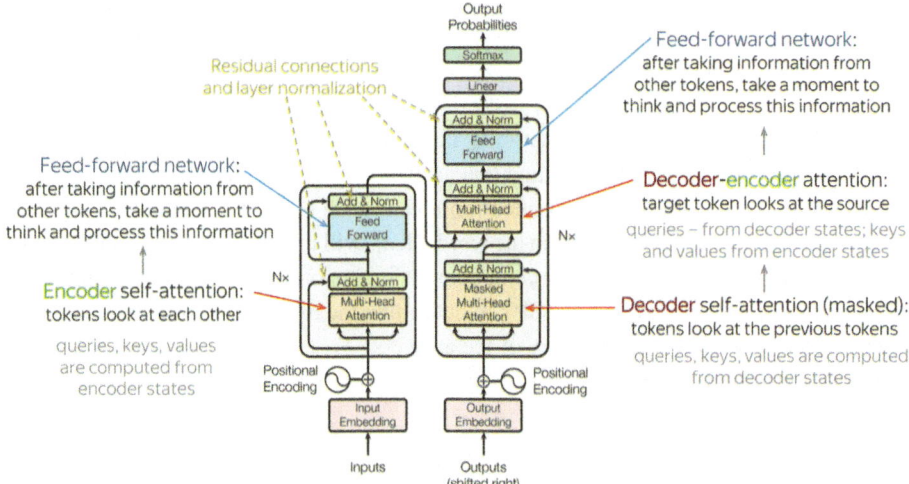

Fig. 4.2 Image depicting the key concepts and functionalities of encoder-only and decoder-only transformer models, as discussed in "Navigating transformers: a comprehensive exploration of encoder-only and decoder-only models, right shift, and beyond" [4]

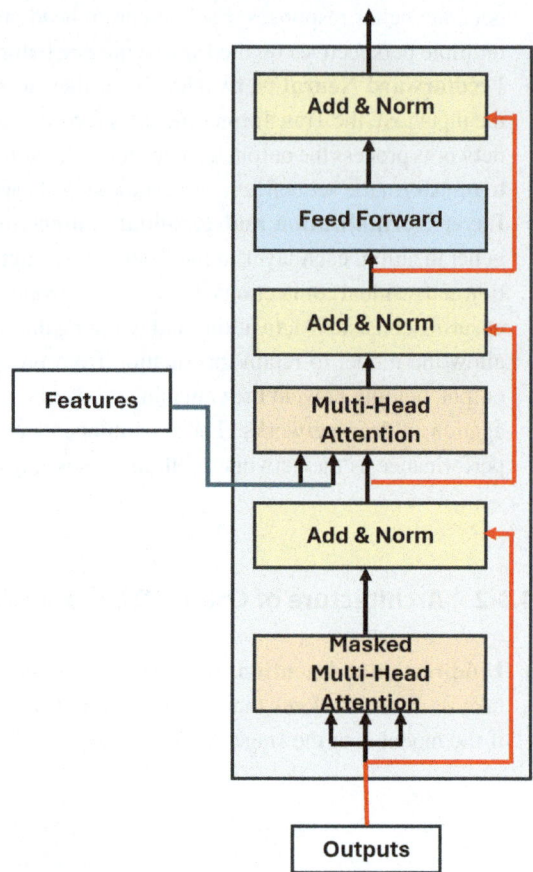

Fig. 4.3 Decoder (reproduced from "Attention is all you need" [3])

from "I" and "love" to decide what to write next. This would mean that each token is produced depending on the preceding tokens available in the context.
- **Self-Attention Mechanism**: Self-attention means each token can attend to every other token in a sequence. For example, in the word, "Ice," will look at "I," "Love," and "cream" to get a better understanding of the context to assign importance to each succeeding word in a sentence. It also enables ChatGPT to determine the relation of certain words next to each other which enhances the flow of the answers.
- **Equation for Self-Attention**:

$$\text{Attention}(Q, K, V) = \text{softmax}\left(\frac{QK^T}{\sqrt{d_k}}\right) V, \qquad (4.4)$$

where Q is the query matrix, K is the key matrix, V is the value matrix, and d_k is the dimension of the key vectors.

4.3 Architecture of ChatGPT

- **Masked Self-Attention**: While in the decoder the self-attention mechanism is slightly altered to exclude the possibility of attending to the future tokens in the input sequence. This is called "masked" self-attention and is very important when dealing with sequential information. In the case of ChatGPT, masked self-attention is applied in such a way that the model predicts from left to right, one token at a time, and thus self-attention is masked to preserve causality. For instance, when forming the "ice cream," the word "cream" cannot use knowledge of "I" or "love." This is under a triangular prism which has a masking effect on the subsequent token.
- **Masking Mechanism**: To make the tokens attend only to the previous tokens and not the future tokens a triangular mask is applied to the self-attention matrix. This mask makes everything above the diagonal as small as possible or negative infinity so that any token cannot look at the future positions, hence autoregressive text generation.

$$\text{Masked Attention}(Q, K, V) = \text{softmax}\left(\frac{QK^T - M}{\sqrt{d_k}}\right)V, \tag{4.5}$$

where Q is the query matrix, K is the key matrix, V is the value matrix, d_k is the dimension of the key vectors, and M is the masking matrix that contains positive infinity ($+\infty$) for positions that should be masked (not attended to) and 0 for positions that can be attended to (Fig. 4.4).

- **Multi-Head Attention (MHA)**: To increase the ability of the model to process a variety of attributes of the input data a modified tool, multi-head attention, is used in ChatGPT. This approach implies having first multiple attention heads, each of these attending in

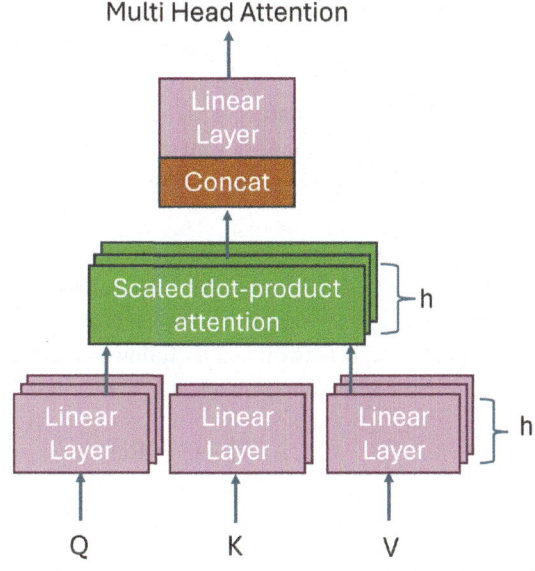

Fig. 4.4 Multi-head Attention (MHA) Taken from "Attention is all you need" [3]

parallel to different parts of the input sequence. For example, one head "is" might be associated with the "I" focus and the other is associated with the "ice cream" focus. MHA uses multiple heads to work as follows:

$$\text{MultiHead}(Q, K, V) = \text{Concat}(\text{head}_1, \text{head}_2, \ldots, \text{head}_h) W^O \quad (4.6)$$

and

$$\text{head}_i = \text{Attention}(Q W_i^Q, K W_i^K, V W_i^V) \quad (4.7)$$

- Q, K, V are the input query, key, and value matrices. - W_i^Q, W_i^K, W_i^V are the parameter matrices for the i-th attention head. - W^O is the output projection matrix after concatenation. - h is the number of attention heads.
- **Benefits of Multi-Head Attention**:

  - **Diverse Perspectives**: Multiple heads allow the model to simultaneously focus on different parts of the sequence, capturing a richer set of relationships.
  - **Enhanced Understanding**: By processing the input from multiple angles, the model gains a more comprehensive understanding of the context.

- **Layered Structure**:
  ChatGPT's architecture comprises multiple stacked layers of Transformer decoders. Each layer refines the representation of the input data, enhancing the performance of subsequent layers. For example, GPT-3 includes 96 such layers.

  - **Stacked Decoder Layers**: In ChatGPT's decoder-only architecture, conversation history serves as the input. The model generates the next token in the sequence based on this context. Stacking decoder layers enables progressively complex feature extraction.
  - **Hierarchical Feature Extraction**: Each layer processes input tokens using self-attention, capturing relevant information and dependencies across the entire sequence to refine context understanding.

- **Feedforward Neural Networks**: After self-attention, tokens are processed through feedforward networks in each layer. These networks apply linear transformations and non-linear activations to capture complex language patterns and relationships. The equation for feedforward networks is as follows where each feedforward layer consists of two linear transformations with a ReLU activation:

$$\text{FFN}(x) = \text{ReLU}(x W_1 + b_1) W_2 + b_2, \quad (4.8)$$

4.3 Architecture of ChatGPT

where

- x is the input to the feedforward network.
- W_1 and W_2 are the weight matrices for the two linear transformations.
- b_1 and b_2 are the bias vectors for the two linear transformations.
- ReLU is the rectified linear unit activation function.

- **Benefits of feedforward neural networks in ChatGPT**:

  - Pattern Recognition: These networks enable the model to learn and represent intricate patterns in language.
  - Increased Expressive Power: Non-linear transformations enhance the model's ability to capture complex relationships.

- **Residual connections and layer normalization**: The decoder embeddings are also connected to each other through residual connections while layer normalization for each layer in the decoder is applied to increase stability in the process. Such mechanisms play a role in increasing the amount of training convergence and, therefore, the quality of the model itself. The equation for layer normalization is

$$\text{LayerNorm}(x) = \frac{x - \mu}{\sigma} \cdot \gamma + \beta \tag{4.9}$$

where:

- x is the input vector.
- μ is the mean of the input vector x.
- σ is the standard deviation of the input vector x.
- γ is a learnable scaling parameter.
- β is a learnable shifting parameter.

- **Positional Encodings**: As mentioned earlier, Transformer models operate over all the input tokens at once and therefore do not learn the temporal dependency of the data. This is however noticed by GPT models through the use of positional encodings that give information about the position of every word in the sequence. This makes it possible for the model to recognize the order of words or the position of the words in a string and make the estimates on that order. For instance, the model is aware of the fact that "cream" appears after "ice" as a result of such encodings. Equation for positional encoding:

$$PE(pos, 2i) = \sin\left(\frac{pos}{10000^{\frac{2i}{d_{model}}}}\right) \tag{4.10}$$

$$PE(pos, 2i+1) = \cos\left(\frac{pos}{10000^{\frac{2i}{d_{model}}}}\right) \quad (4.11)$$

where

- $PE(pos, 2i)$ represents the positional encoding at position *pos* for even indices.
- $PE(pos, 2i+1)$ represents the positional encoding at position *pos* for odd indices.
- *pos* is the position of the word in the sequence.
- *i* is the dimension index.
- d_{model} is the dimensionality of the embedding.

- Benefits of positional encodings in ChatGPT:

    - Sequential Order: Some positional encoding techniques are used to make the model aware of the order or tokens which is very important for context.

4.3.3 Reinforcement Learning from Human Feedback (RLHF)

Human-in-the-loop reinforcement learning from human feedback (RLHF) is a technique employed to further refine machine models, especially in NLP and conversational AI, which is dictated by the feedback of human evaluators [5]. It improves the quality of the responses that are produced by adjusting them to human parity and standards.

Here's a detailed explanation of the RLHF process:

1. **Initial Model Training**
   The procedure starts with a pre-trained natural language processing model like GPT. This first training is usually conducted relying on large-scale unsupervised learning on the multitype text data. The general language model, grammar, consistencies, and semantic relations are learned by the model, but it may not correspond well with the user's preferences and the current conversational objectives.
2. **Supervised Fine-Tuning**
   To improve the model, it undergoes supervised fine-tuning using a dataset curated by human AI trainers. These trainers simulate real-life dialogs by providing the model with inputs and corresponding appropriate outputs. This dataset serves as a reference for adjusting the model's behavior, enabling it to generate responses that more closely align with human expectations.

4.3 Architecture of ChatGPT

Supervised Fine-Tuning Process:

(a) **Data Collection**: Some of the utterances are generated by human trainers, while others are artificially created through automated exercises. The responses are also ranked based on quality.

(b) **Training**: This dataset is used to train the model, enabling it to produce responses that closely resemble those of the trainers.

3. **Creating a Reward Model**

Another crucial step in RLHF is evaluating the quality of the model's responses using a reward model. In this process, human evaluators review the model's responses and rank them in order of preference. These rankings are then used to define a reward function, which assigns scores to the generated responses, measuring their quality, coherence, and alignment with human preferences.

Reward Model Construction:

(a) **Ranking Responses**: For a given input prompt, human evaluators assign ranks to multiple responses.

(b) **Training the Reward Model**: These rankings are used to train a reward model, where features such as relevance, informativeness, and fluency are input to estimate the quality of a response.

4. **Reinforcement Learning Optimization**

The straightforward use of the reward model is followed by its integration into a reinforcement learning setting for fine-tuning the language model. Indeed, this entails finding the best-fitting response generation policy that yields the expected reward according to the reward model. For this purpose, we use the proven method called Proximal Policy Optimization (PPO). PPO is a form of reinforcement learning that seeks to adjust the policy in a manner that allows the agent to perform various tasks while minimizing the chances of reverting to previous suboptimal behaviors while at the same time ensuring that the agent is not exploited when new strategies are being tested. We have learned what PPO seeks to achieve, which is to enhance the policy with updates that need not be far from the previous policy.

The PPO objective function is defined as

$$L^{CLIP}(\theta) = \mathbb{E}_t \left[\min \left(r_t(\theta) \hat{A}_t, \text{clip}(r_t(\theta), 1 - \epsilon, 1 + \epsilon) \hat{A}_t \right) \right], \quad (4.12)$$

where

- The probability ratio is given by

$$r_t(\theta) = \frac{\pi_\theta(a_t|s_t)}{\pi_{\theta_{old}}(a_t|s_t)}$$

which is the ratio of the new policy π_θ to the old policy $\pi_{\theta_{old}}$.
- $\hat{A}_t$ is the advantage function, which measures how much better the action taken is compared to the average action in that state.
- ϵ is a clipping factor that limits the size of the policy update to prevent large changes that might degrade the policy.

5. **Iterative Refinement**
Iterative Refinement means the symbol set is a dynamic model that is refined constantly with new data produced by the reward model fed into it periodically. This cycle of operation goes on until the performance of the model meets the objectives and is closely in line with the expectations of human evaluators. Iterative Refinement Steps are as follows:

   (a) Generate Responses: The model responds to a set of questions with answers.
   (b) Evaluate Responses: The reward model then rates these responses in terms of the quality of the response that they will provide.
   (c) Update Model: Other than that, the scores are used in tuning the model in a bid to increase the level of response quality.

4.4 Pre-training and Fine-Tuning in ChatGPT

4.4.1 Pre-training

There is some recent work where learned models are pre-trained in a way that they are required to see lots of text data during this phase, so that they get a good sense of language in general. During this phase, the model is exposed to more formal grammar and syntax of language as well as its more routine usage. Before Phase 2, the model applies an unsupervised learning method in which it is to predict the subsequent word in a sequence with the prior words as input. This process is usually done using a large text dataset, such as text messages. The objective function applied during the pre-training phase may be the cross-entropy, which estimates how close the output probability distribution is to the actual distribution of the next word in the sequence.

For a given sequence $\{w_1, w_2, \ldots, w_{n-1}\}$, the model learns to predict w_n using

$$P(w_t \mid w_1, w_2, \ldots, w_{t-1}), \tag{4.13}$$

4.4 Pre-training and Fine-Tuning in ChatGPT

where $P(w_t \mid w_1, w_2, \ldots, w_{t-1})$ refers to the probability of occurrence of the word w_t based on the preceding words $w_1, w_2, \ldots, w_{t-1}$.

4.4.2 Fine-Tuning: Increase/Decrease Repetition Depending on the Tasks

In post-training, which is known as fine-tuning, the model is made to learn specific tasks and is fine-tuned to domains. The first one involves supervised learning and includes focusing on the specific domain data and interacting with the users.

- Domain Adaptation:
  The fine-tuning, therefore, helps to adjust the model's knowledge toward application or domain. For instance, if the model is meant for customer service, then it will be fine-tuned on customer service conversations. This enables the model to capture the peculiarities and specifications of the particular domain of interest.
- User Interaction Guidance:
  Second, fine-tuning consists of guiding user interactions within the target domain. This step further guarantees that ChatGPT is friendly and appropriate to answer in any given context while also being responsible. Tools like RLHF are employed to encourage the right behaviors and avoid or at least discourage damaging responses.
  Let the task under consideration be represented by a task-specific loss function, and it is the task of fine-tuning to minimize it. For example, in classification tasks, the loss function could be

$$L(\mathbf{y}, \hat{\mathbf{y}}) = \frac{1}{n} \sum_{i=1}^{n} \ell(y_i, \hat{y}_i), \tag{4.14}$$

where

- y_i represents the true label for instance i,
- $\hat{y}_i$ are the predicted labels obtained from the model,
- n is the total number of instances, and
- ℓ is the loss function, which measures the difference between the true and predicted labels.

4.4.3 Continuous Learning and Iterative Improvement

This principle emphasizes the continuous learning and improvement necessary to sustain and optimize ChatGPT. It ensures that the model remains effective in interpreting and generating text as language dynamics evolve and new knowledge is incorporated.

- **Continuous Learning**: In ChatGPT, learning is an ongoing process, with the model being updated or fine-tuned periodically as new training data becomes available. This process ensures the model stays aligned with changes in language, emerging topics, and evolving user needs.

  - **Updating with New Data**: ChatGPT is continuously trained with new data that includes the latest discussions, topics, and changes in information. This helps the model remain up to date and relevant.
  - **Adapting to Evolving Patterns**: Alongside new data, ChatGPT is updated to reflect shifts in language and trends, improving the generativeness and relevance of its responses.
  - **Mechanism of Update**: The update process is usually performed using gradient descent optimization—a tool that allows changing the model parameters to reduce the value of the loss function. The update rule for model parameters θ can be expressed as

  $$\theta \leftarrow \theta - \eta \nabla_\theta L(\theta), \tag{4.15}$$

  where

    * θ represents the model parameters,
    * η is the learning rate, which controls the size of the update step,
    * $\nabla_\theta L(\theta)$ denotes the gradient of the loss function L with respect to the model parameters.

- **Iterative Improvement**:
  The process of progressive enhancement involves continuously improving ChatGPT by refining metrics, addressing user feedback, and analyzing performance. This iterative process enhances the quality of the model, increasing its precision, relevance, and responsiveness.

  - **Performance Evaluation**: Adjustments are periodically made based on evaluations of how effectively ChatGPT provides contextually relevant and accurate responses. Key metrics such as accuracy, coherence, and user satisfaction are used to quantify performance.
  - **Feedback Integration**: User interactions and feedback offer valuable insights into the model's potential weaknesses. This feedback is then used to refine and improve the model in specific areas of focus.
  - **Hyperparameter Tuning and Architecture Modification**: Based on model assessments, adjustments may be made to machine learning algorithms, including tuning hyperparameters like learning rates or batch sizes, or modifying the model's architecture to enhance its efficiency.

4.4 Pre-training and Fine-Tuning in ChatGPT

- **Mathematical Representation of Updates**: The iterative refinement process can be seen through gradient-based optimization techniques as well. The above chart is self-explanatory. The parameter updates during iterative improvement are applied as follows:

$$\theta^{(t+1)} = \theta^{(t)} - \eta \nabla_\theta L(\theta^{(t)}), \qquad (4.16)$$

where

* $\theta^{(t)}$ are the model parameters at iteration t,
* $\theta^{(t+1)}$ are the updated model parameters at the next iteration,
* η is the learning rate,
* $\nabla_\theta L(\theta^{(t)})$ is the gradient of the loss function concerning the parameters at iteration t.

4.5 Contextual Embeddings in ChatGPT

These passing contexts or frames are the contextual embeddings that make a significant difference in the way ChatGPT generates contextually relevant and coherent answers. Here's an in-depth look at their role and functionality.

4.5.1 Role of Contextual Embeddings

Contextual embeddings are vectors that capture the meaning of words or tokens, which vary depending on the surrounding text. This allows language to be understood in context. This contrasts with traditional meaning representations, where words are represented by fixed vectors, whereas in contextual embeddings, words are represented by vectors that depend on their position within a sentence or document.

Key aspects of contextual embeddings include:

- Dynamic Representation: Then in ChatGPT, the position and the words that surround the word or the token are taken into consideration to capture the difference between words or tokens. One advantage of the dynamic nature is that it enables capturing of semantically related forms of the same word depending on the use of the word in context.
- Contextualized Understanding: The model produces the embeddings that correspond to the semantic and syntactic function of words in certain context. This allows ChatGPT to generate outputs which are sensible and in relation to the input, thus keeping in touch with the required exchange.

4.5.2 Generating Contextual Embeddings

The process of generating contextual embeddings involves several steps:

- **Input Tokenization**: The input text is preprocessed using a chosen tokenizer, which breaks the text down into units called tokens. Each token is then assigned an initial vector. For example, the sentence "It is amazing" would be tokenized into the tokens "It," "is," and "amazing."
- **Embedding Transformation**: The initial embeddings of the input sequence pass through several layers of the Transformer model. Each layer contains sub-components, including self-attention, which recalculates the embeddings based on the input from other tokens in the sequence.
- **Self-Attention Mechanism**: This mechanism computes attention scores for each token in relation to other tokens, allowing the model to focus on specific components of the input sequence. The attention score for a token t_i in relation to token t_j can be expressed as

$$s_{ij} = \frac{\exp\left(\text{score}(t_i, t_j)\right)}{\sum_{k=1}^{N} \exp\left(\text{score}(t_i, t_k)\right)}, \qquad (4.17)$$

where

- s_{ij} is the attention score for token t_i in relation to token t_j,
- $\text{score}(t_i, t_j)$ is the compatibility score between tokens t_i and t_j,
- N is the total number of tokens in the sequence.

- **Enhanced Coherence**: Long context helps the model understand the context of the text it is responding to or follow coherent threads in longer passages. Since the model has the context of the words within and between sentences, it can generate outputs that logically reflect both the content of the text and the overall situation.
- **Improved Relevance**: This involves generating embeddings that incorporate context to create responses highly relevant to the input query. This reduces ambiguities in the model's responses and enhances its ability to produce meaningful and contextually appropriate answers.
- **Flexibility in Language Understanding**: The model can better understand language through contextual embeddings by accounting for various linguistic phenomena, such as polysemy and syntactic variations, that depend on the context.
- **Response Generation in ChatGPT**: Another complex aspect of response generation in ChatGPT is the need to develop sophisticated techniques to address tasks and generate coherent, contextually sound prompts. Here's a detailed breakdown of how this process unfolds:

- **Autoregressive Generation**:
  Among all the mechanisms of the ChatGPT model, autoregressive generation is one of the most important. The process begins with encoding each word using the start-of-sequence token, denoted as <sos>, which signals the beginning of the generation sequence. The response generation proceeds as follows:

  * **Initialization**: Thus, the model begins the generation with the <sos> token and generates the first word, making this prediction out of the primary context:

  $$e_{<sos>} \text{ denotes the mapping of the start-of-sequence token.} \quad (4.18)$$

  * **Sequential Prediction**: The subsequent word generation can be represented as
  $$c_t = f(h_{t-1}, e_{w_{t-1}}), \quad (4.19)$$
  where c_{t-1} is the contextual embedding including the previously generated words, h_{t-1} is the hidden state from the previous time step, and $e_{w_{t-1}}$ is the embedding of the previously generated word.

- **Probability Distribution**: At each step in the sequence, ChatGPT calculates a probability distribution over the entire vocabulary to choose the next word:

  * **Probability Calculation**: Given the current contextual embedding, the model computes a probability distribution for all possible next words as follows:
  $$P(w_t|c_t) = \text{softmax}(W_{\text{out}}h_t + b_{\text{out}}), \quad (4.20)$$

  where
  $P(w_t|c_t)$ is the probability of the next word w_t given the contextual embedding c_t,
  W_{out} is the output weight matrix,
  b_{out} is the output bias, and
  h_t is the hidden state at time t.

- **Word Selection**:
  Several methods can be used to select the next word from the probability distribution:

  * **Greedy Decoding**: Selects the word with the highest probability. This method is deterministic but can be less diverse.
  * **Random Sampling**: Chooses a word randomly based on the computed probabilities, introducing variability into the responses.

* **Top-k Sampling** Selects from the top-k highest probability words, improving diversity while controlling randomness.
* **Nucleus Sampling**: Chooses from words whose cumulative probability exceeds a certain threshold $p_{\text{threshold}}$, allowing for more flexible and diverse outputs.

- **Response Construction**:
  When a word has been chosen, it adds to the end of the response chain. The contextual embeddings are updated to reflect this addition, and the process continues iteratively:

  – Embedding Update: The new word is added to the sequence, and contextual embeddings are recalculated as follows:
  $$c_t = \text{Update}(c_{t-1}, w_t), \tag{4.21}$$
  where
  * c_t is the updated contextual embedding,
  * c_{t-1} is the previous contextual embedding,
  * w_t is the newly added word.

  – Termination: This process repeats until the model generates an end-of-sequence token `<eos>` or reaches a maximum sequence length L:
  $$\text{If } w_t = \text{<eos> or } t = L, \text{ terminate.} \tag{4.22}$$

- **Unified Architecture and Policy**:
  The process of response generation is contained within the ChatGPT framework. In the case of ChatGPT, the term "policy" is not an object but a set of learned weights and parameters of the model. These weights reflect the model's knowledge of language patterns, contexts, and behaviors acquired during training.

  – Policy Influence: The policy determines the kind of words which are used and how responses to queries should be formulated. It follows the statistical language used in training as opposed to comprehension or even designing.

4.5.3 Example: Generating a Response for the Input "What Are the Benefits of Regular Exercise?"

1. **Initialization** Input: "What are the benefits of regular exercise?"
   Start-of-Sequence Token: `<sos>`
   The generation process begins with the start-of-sequence token `<sos>` to signify the beginning of the response sequence.
2. **Generating the First Word** Contextual Embedding: The model computes the initial contextual embedding based on the input question.

$$e_{\text{sos}} = \text{ComputeEmbedding(Input)} \tag{4.23}$$

Prediction: The model predicts the first word in the response sequence.

$$p_1 = \text{Softmax}(W_{\text{out}} \cdot h_1 + b_{\text{out}}). \tag{4.24}$$

Suppose the predicted probability distribution indicates high likelihood for the word "Regular."
Word Selection: The model selects "Regular" using one of the decoding strategies. For this example, let's use greedy decoding.

$$\text{Token}_{\text{next}} = \arg\max(p_1) = \text{"Regular"}. \tag{4.25}$$

Update Context: The response sequence is updated, and the contextual embedding is recalculated.

$$c_1 = \text{UpdateEmbedding}(e_{\text{sos}}, \text{"Regular"}). \tag{4.26}$$

3. **Generating the Second Word** Contextual Embedding Update: The new contextual embedding reflects the addition of "Regular."

$$e_2 = \text{ComputeEmbedding("Regular")}. \tag{4.27}$$

Prediction: The model predicts the next word based on the updated context.

$$p_2 = \text{Softmax}(W_{\text{out}} \cdot h_2 + b_{\text{out}}). \tag{4.28}$$

Suppose the predicted probabilities favor the word "exercise."
Word Selection: The model selects "exercise" using greedy decoding.

$$\text{Token}_{\text{next}} = \arg\max(p_2) = \text{"exercise"}. \tag{4.29}$$

Update Context: The response sequence is updated to "Regular exercise," and the contextual embedding is recalculated.

$$c_2 = \text{UpdateEmbedding}(e_2, \text{"exercise"}). \tag{4.30}$$

4. **Generating the Third Word** Contextual Embedding Update: The new embedding reflects "Regular exercise."

$$e_3 = \text{ComputeEmbedding}(\text{"Regular exercise"}). \quad (4.31)$$

Prediction: The model predicts the next word.

$$p_3 = \text{Softmax}(W_{\text{out}} \cdot h_3 + b_{\text{out}}). \quad (4.32)$$

Suppose "improves" is highly probable.
Word Selection: "Improves" is selected.

$$\text{Token}_{\text{next}} = \arg\max(p_3) = \text{"improves"}. \quad (4.33)$$

Update Context: The sequence now reads "Regular exercise improves," and the context is updated.

$$c_3 = \text{UpdateEmbedding}(e_3, \text{"improves"}). \quad (4.34)$$

5. **Continue Generation** The process continues with similar steps: Contextual Embedding Update: Updates reflect the growing sequence.

$$e_4 = \text{ComputeEmbedding}(\text{"Regular exercise improves"}). \quad (4.35)$$

Prediction: Predicts the next word.

$$p_4 = \text{Softmax}(W_{\text{out}} \cdot h_4 + b_{\text{out}}). \quad (4.36)$$

Let's say "overall" is predicted with high probability.
Word Selection: "Overall" is selected.

$$\text{Token}_{\text{next}} = \arg\max(p_4) = \text{"overall"}. \quad (4.37)$$

Update Context: The sequence updates to "Regular exercise improves overall," and the context is recalculated.

$$c_4 = \text{UpdateEmbedding}(e_4, \text{"overall"}). \quad (4.38)$$

6. **Termination**: This process continues until the model generates an end-of-sequence token <eos> or reaches a predefined maximum length. For instance, the final sequence might be Complete Response: "Regular exercise improves overall health and well-being."

4.6 Handling Biases and Ethical Considerations

4.6.1 Addressing Biases in Language Models

ChatGPT and other language models are trained on billions of texts collected from the Internet. As we will see, this data is not free from biases, and as a result, the model may also learn these biases. Such biases may include racism, sexism, and political prejudice, which could be reflected in the model's responses. Persistent biases can influence how the AI interacts with users, potentially producing outputs that are vulgar, undesirable, or biased.

4.6.2 Awareness of the Consequences of Biases

When developing AI solutions, there are several potential pitfalls, and unchecked biases in AI models can be particularly dangerous. They can perpetuate prejudice, provide incorrect information, and may not be inclusive or friendly to all individuals. Therefore, it is crucial to address these biases to ensure that interactions between humans and intelligent machines are virtuous and respectful.

4.6.3 Strategies Adopted by OpenAI to Eliminate Biases

1. **Fine-Tuning with Human Supervision**
   After the pre-training phase, OpenAI continues to improve ChatGPT through fine-tuning. This process involves human reviewers who assess the model's outputs using a standard method, providing their feedback without being influenced by external factors. Reviewers focus on model outputs for various input examples, comparing them to recommended responses and offering their impressions to help enhance the model's performance. This iterative process enables the model to adjust its approach, though it requires significant resources and extended deployment times.
   Example: If a user asks, "What are the advantages of practicing meditation?" The user can supervise the response prompts and make changes according to there needs.
2. **Regular Updates to Guidelines**
   Some of the guidelines integrated into Solimon are updated in response to societal changes, while others are adjusted based on user feedback. Additionally, OpenAI holds more frequent meetings to address questions and clarify doubts, making the model more receptive to diverse opinions. However, reaching a consensus on rule changes can be complex.
   Example: Upon discovering that recent research provides new insights on how certain topics should be addressed, adjustments are made to ensure responses are more "politically correct."

3. **Transparency** OpenAI values transparency in its processes, goals, and strategies, as well as the honesty regarding the imperfections of its AI models. Public suggestions, recommendations, and feedback are welcomed to help enhance the proper use of the technology. However, it is constrained by the challenges of achieving precise and perfect AI systems, as well as by concerns over user privacy.
   Example: OpenAI may publish separate papers on how biases are detected and addressed in ChatGPT, providing insight into the efforts to ensure fairness.
4. **Research and Development**
   OpenAI's ChatGPT focuses on continual learning aimed at reducing both discriminatory and implicit biases. One key aspect of this effort is defining how the model should respond to sensitive topics that may be considered taboo in certain societies, as well as determining which behaviors are more appropriate in specific cultural or national contexts. These research initiatives aim to broaden the model's applicability in handling contentious and sensitive issues in a more equitable manner.
   Example: The model may focus on mitigating bias when making predictions related to gender-based questions, with the goal of developing a fairer model.
5. **Customization and User Feedback**
   OpenAI plans to introduce a set of options that allow users to control ChatGPT's actions according to specific societal norms. This kind of personalization aims to tailor the model to individual users, improving the results and gathering data on the model's effectiveness. However, customization also presents challenges, such as determining what is acceptable in a user-modified environment and preventing users from making adjustments that could lead to unethical outcomes.
   Example: A user may wish to adjust the tone of ChatGPT to avoid discussing certain topics. This feedback can contribute to the fine-tuning of the model, improving its adaptation to the user's specific needs.

4.6.4 Challenges and Trade-Offs

Addressing biases in AI is challenging, and each approach comes with its own trade-offs. Measures to minimize bias often result in increased costs and can negatively impact the models. Similarly, the fundamental issue of defining and obtaining unbiased data remains unresolved. OpenAI continues to work on reducing bias, acknowledging that while improvements can be made to bias detection systems, completely eliminating biases is not yet achievable.

4.6.5 Motivation for Growth

In order to overcome challenges, OpenAI will continue striving to further reduce bias and adopt the most suitable ethical practices in the development of artificial intelligence. Valuable lessons can be learned through collaborative discussions, exchanging feedback, and working to develop AI systems that align with diverse perspectives and values.

4.7 Strengths and Limitations of ChatGPT

4.7.1 Strengths of ChatGPT

1. **Conversational Prowess and Adaptability**: The major strength of the ChatGPT is its ability to converse contextually on a wide range of issues and in a coherent manner. That's why it is remarkably useful for different tasks, including customer support, education, game-based learning, and creating interactive narratives. For example, if ChatGPT is implemented in a customer support operation, it will be possible for the artificial intelligence system to comprehend customer queries and offer precise responses that are relevant to specific circumstances.
   Example: Versatile Response Generation
   Scenario: A user interacts with ChatGPT in a conversation covering general and casual conversation and even technical inquiries.
   User: I want to know the positive effects of exercise "Please, can you enumerate the advantages that one can get from exercising?"
   ChatGPT Response: "There are several advantages of exercising including a better heart condition, mood, and performance of the brain. It also assists in weight control and increase of energy levels.
2. **Fine-Tuning Process**: ChatGPT's ability to generate safer and more useful responses is enhanced through fine-tuning by OpenAI, which incorporates human input into the model. This process helps the model align more closely with human values and ethical perspectives. For example, if a user provides feedback indicating that certain responses are inappropriate, the model can be adjusted to avoid generating similar incorrect responses in the future, making it more appropriate overall.
   Example: Reducing Harmful Outputs
   Scenario: When the model is found to be giving out wrong responses and inconsiderate content, OpenAI performs the process of fine-tuning.
   Original Response: "Some issues are best described as sensitive since the perception they get is often skewed."
   Revised Response (after fine-tuning): "Discussions concerning sensitive issues should be conducted properly without fueling hatred and/or enraging the other party."

3. **Iterative Development**:
   ChatGPT has an advantage due to its continuous updates and improvements in a cyclical development process. These updates are driven by feedback from users, including clients, as well as advancements in AI research. As a result, each new iteration of the model, from GPT-1 to GPT-4, reflects improvements in the realism and relevance of generated prompts. This ongoing process allows the model to adapt to users' needs and address emerging societal issues.
   Example: Changes made from GPT-3 to GPT-4.
   Scenario: The users' responses of GPT-3 and GPT-4 on a query are presented where GPT-4 comprehends a more sophisticated completion.
   GPT-3 Response: "Controversial theories; the theory of relativity is associated with Einstein and his work on time travel."
   GPT-4 Response: This is according to the theory of relativity by Albert Einstein that spelled new dimensions of space and time though it has nothing to do with what is seen in science fiction about time travel.

4.7.2 Limitations of ChatGPT

1. **Lack of World Knowledge**: Despite its capabilities, ChatGPT's understanding of the world is limited to the patterns it has been trained on and the information available up to its last update. For example, when answering questions related to recent news or current events, ChatGPT may provide outdated or incorrect information, as it is unaware of real-time developments.
   Example: Current Events.
   Scenario: A user asks about a particular event that occurred in the recent past.
   User: "What news is there about the 2024 Summer Olympic Games?"
   ChatGPT Response: "Unfortunately I can only rely on the information up to [training cutoff date], For the most recent updates please refer to other sources such as newspapers."
2. **Biases**: ChatGPT can produce responses based on the bias present in the data used while training the model. Some of these biases, however, may still exist even at the end of the fine-tuning phase that is usually conducted to minimize them. For example, if the training data contains racist or sexist comments, we'll still see that ChatGPT makes racist or sexist remarks due to those inputs despite efforts toward preventing it.
   Example: Cultural Sensitivity **Scenario**: Cultural practices: ChatGPT will further answer this question.
   User: Some of the most popular customs that are observed in different cultures include:
   ChatGPT Response: "Some traditions that they have are very basic and simple such as special kinds of diets and celebrations that may differ greatly."

3. **Inappropriate or Unsafe Outputs**: Despite this, there are times when ChatGPT can generate outputs which are unsafe, unfamiliar, vulgar, or abusive. These outputs are not precise, they are not the outcomes that have been consciously aimed at, but the results of the training of the model. For instance, the response might contain information that is prohibited or reveal information because the model learns from all the data.
   Example: Handling Sensitive Queries **Scenario**: There is such a question from a user: what is your opinion about... cruel, aggressive topic?
   User: "Could you share with me a joke related to politics or any topic that is better not to be touched?"
   ChatGPT Response: "I want only good things on my timeline; how about a joke that will not lead to a debate or offend anyone?"
4. **Absence of Common Sense or Deep Understanding**: Moreover, it has been established that ChatGPT fails to have actual intelligence or even basic cause-and-effect thinking. It produces responses that are learned through patterns, and not intelligence. This, in turn, can lead to responses that though may appear rational are, in fact, irrational and/or erroneous. For instance, ChatGPT can spit out a seemingly realistic answer bored of a profound idea because it does not comprehend context in the manner of a human.
   Example: Practical Advice
   Scenario: A user wants to get a recommendation on what to do in a given task that he or she encounters daily.
   User: "There are numerous works that pose the question, 'How can a small apartment be arranged?'"
   ChatGPT Response: "Here are some tips you may try, such as using multi-purpose furniture and making use of storage space."
5. **Inability to Fact-Check**: ChatGPT is unable to validate the information it provides or verify the accuracy of its responses. It generates answers based on patterns learned from its training data, which means it can produce information that sounds plausible but may be incorrect or misleading. For example, when asked about a specific historical event or scientific fact, ChatGPT may give an inaccurate answer since it cannot cross-check the information in real time.
   Example: Historical Facts
   Scenario: A user wants to know more about a particular figure in history.
   User: "Which woman was the first to be awarded the Nobel Prize?"
   ChatGPT Response: "The first woman to be awarded the Nobel Prize was Marie Curie in 1903—for Physics."

4.8 Related Studies

Here are some papers and their contributions to the field:

1. **DistilBERT: A simplified version of BERT**
   "A Primer for BERT" by its authors including Viet Nguyen, Christopher Ward, Graham Neubig, and Ilya Sutskever.
   This paper by Sanh et al. [6] introduces DistilBERT—Mobile & Efficient BERT, which is a small version of the BERT model that loses much of BERT's ability but at a much smaller cost and with much lower computational requirements. Using knowledge distillation, the authors trained a lightweight alternative model, "student," that mimics the behavior of the large, 12-layer "teacher" model, BERT, but with fewer parameters. Concretely, DistilBERT preserves 97% of BERT's linguistic comprehension efficiency but is 60% smaller and 60% swifter since the transformer layers are limited to six. This approach permits DistilBERT to offer extremely high results on other NLP tasks, including text classification and named entity recognition, and therefore it is more useful for environments with limited computational ability.

2. **Scaling Laws for Neural Language Models** (Kaplan et al., 2020) [7]
   This paper provides an in-depth analysis of how increasing the scale of neural language models affects their performance. Kaplan et al. demonstrate that larger models, like GPT-3, and show improved performance on a variety of tasks as the amount of data, model size, and computational resources increase. They establish empirical scaling laws that predict how model performance improves with additional resources, highlighting that larger models tend to generalize better and achieve higher performance on benchmarks.

3. **The Power of Scale for Parameter-Efficient Prompt Tuning** (Lester et al., 2021 [8])
   Lester and colleagues explore parameter-efficient techniques for adapting large language models to specific tasks without requiring extensive retraining. They introduce prompt tuning, which involves fine-tuning a model's prompts to perform various tasks effectively. The study shows that this approach allows models like GPT-3 to achieve task-specific performance improvements while minimizing the number of parameters that need to be trained, thus making the fine-tuning process more efficient.

4. **In-Context Learning and Induction Heads** (Olsson & Steedman, 2022 [9])
   This paper investigates the in-context learning capabilities of large language models, focusing on how they can perform tasks by leveraging contextual information provided during inference. The authors examine the role of induction heads, a specific type of attention mechanism, in facilitating this capability. Their findings suggest that induction heads play a crucial role in allowing models like GPT-3 to understand and generate responses based on contextual cues, without the need for extensive pre-training.

5. **Improving Language Understanding by Generative Pre-Training** (Radford et al., 2022 [10])
   Radford and team explore advancements in language understanding achieved through generative pre-training techniques. They present methods that enhance the performance of large language models on a wide range of NLP tasks by leveraging pre-training on large datasets. The paper emphasizes how these techniques allow models like GPT-3

to develop a deeper understanding of language patterns and contextual relationships, leading to improved task performance.

4.9 Summary

Altogether, this layout of ChatGPT implies a step up in trends of natural language processing and artificial intelligence. The use of the GPT model together with proper pre-training and fine-tuning of the model makes it refill human language in terms of text production across different fields. However, like any AI system, ChatGPT has certain limitations: it has inherent bias, has the capability of providing rude/undesirable responses, and does not have real-time fact-checking or deep understanding. Thus, based on OpenAI's work, it can be highlighted that ongoing research and addressing the essential problems with the help of feedback from users showcases the relevance of responsible AI development. As the new technologies of AI continue to develop, people like ChatGPT will be valuable in introducing a wide range of opportunities and potentials for AI systems as well as the possibilities of risks, unreliability, and ineffectiveness inherent in new technologies of AI.

4.10 Multiple-choice Questions

In this section, you'll find a series of multiple-choice questions designed to test your understanding of key concepts in generative AI. Choose the correct answer for each question.

1. Which of the following innovations was introduced with GPT-1?
   (A) Few-Shot Learning
   (B) Multimodal Capabilities
   (C) Unidirectional Transformer Architecture
   (D) Zero-Shot Learning

2. What was a major advancement of GPT-2 over its predecessor, GPT-1?
   (A) Introduction of multimodal learning
   (B) Increase in the number of parameters from 175 billion to 1.5 billion
   (C) Implementation of unsupervised learning on a diverse dataset
   (D) Development of ethical considerations and content filtering

3. Which GPT model demonstrated the ability to perform tasks with very few examples, showcasing its generalization capabilities?
   (A) GPT-1
   (B) GPT-2
   (C) GPT-3
   (D) GPT-4

4. What key feature of GPT-4 focuses on addressing ethical concerns and reducing the potential for harmful content?

   (A) Few-Shot Learning
   (B) Enhanced Fine-Tuning
   (C) Multimodal Capabilities
   (D) Increased Model Size

5. As of 2024, what is a prominent feature of GPT-4 and later models that distinguishes them from earlier versions?

   (A) The use of recurrent neural networks
   (B) Integration of text with other data types like images and audio
   (C) A model size reduction to 1.5 billion parameters
   (D) Focus solely on unsupervised learning

6. What innovation does the self-attention mechanism in the Transformer model use to capture long-range dependencies?

   (A) Unidirectional Attention
   (B) Multi-Head Attention
   (C) Contextual Encoding
   (D) Residual Connections

7. Which component of the Transformer architecture in ChatGPT allows the model to process entire sequences in parallel rather than sequentially?

   (A) Positional Encodings
   (B) Self-Attention Mechanism
   (C) Feedforward Neural Networks
   (D) Masked Attention

8. In the decoder-only structure of ChatGPT, which mechanism ensures that each token only considers preceding tokens and not future ones?

   (A) Self-Attention Mechanism
   (B) Multi-Head Attention
   (C) Masked Self-Attention
   (D) Feedforward Networks

9. What does the Reinforcement Learning from Human Feedback (RLHF) process utilize to improve the model's responses after initial training?

   (A) Supervised Fine-Tuning
   (B) Unsupervised Learning
   (C) Reward Model and Proximal Policy Optimization (PPO)
   (D) Layer Normalization and Residual Connections

4.10 Multiple-choice Questions

10. Which equation represents the mechanism used in the self-attention layer to compute attention scores in the Transformer architecture?

    A) $\text{Attention}(Q, K, V) = \text{softmax}(QK^\top)V$
    B) $\text{Attention}(Q, K, V) = \text{softmax}\left(\frac{QK^\top}{\sqrt{d_k}}\right)V$
    C) $\text{Attention}(Q, K, V) = Q + K + V$
    D) $\text{Attention}(Q, K, V) = \text{sigmoid}(QK^\top)V$

11. During the pre-training phase of ChatGPT, which objective function is commonly used to measure the model's performance?

    (A) Mean Squared Error
    (B) Cross-Entropy Loss
    (C) Hinge Loss
    (D) Kullback–Leibler Divergence

12. In the fine-tuning phase of ChatGPT, which technique is employed to align the model's responses with human preferences?

    (A) Dropout Regularization
    (B) Reinforcement Learning from Human Feedback (RLHF)
    (C) Contrastive Divergence
    (D) Batch Normalization

13. What is the primary purpose of positional encodings in the Transformer architecture used by ChatGPT?

    (A) To provide static word embeddings
    (B) To capture the sequential order of tokens
    (C) To initialize hidden state vectors
    (D) To prevent overfitting during training

14. In the context of ChatGPT's response generation, what does the term "masked self-attention" refer to?

    (A) Preventing future tokens from influencing the current token's prediction
    (B) Concealing input tokens to enhance privacy
    (C) Using attention masks to highlight important tokens
    (D) Adjusting attention weights dynamically during training

15. What is the key difference between greedy decoding and top-k sampling in response generation?

    (A) Greedy decoding introduces randomness, while top-k sampling does not
    (B) Top-k sampling selects the most probable word, while greedy decoding selects randomly

(C) Greedy decoding always chooses the word with the highest probability, whereas top-k sampling selects from the top-k highest probability words

(D) Top-k sampling is deterministic, while greedy decoding is probabilistic

4.11 Answers

Below are the answers to the multiple-choice questions from the previous section:

1. (C) Unidirectional Transformer Architecture
2. (C) Implementation of unsupervised learning on a diverse dataset
3. (C) GPT-3
4. (B) Enhanced Fine-Tuning
5. (B) Integration of text with other data types like images and audio
6. (B) Multi-Head Attention
7. (B) Self-Attention Mechanism
8. (C) Masked Self-Attention
9. (C) Reward Model and Proximal Policy Optimization (PPO)
10. (B) $\text{Attention}(Q, K, V) = \text{softmax}\left(\frac{QK^\top}{\sqrt{d_k}}\right) V$
11. (B) Cross-Entropy Loss
12. (B) Reinforcement Learning from Human Feedback (RLHF)
13. (B) To capture the sequential order of tokens
14. (A) Preventing future tokens from influencing the current token's prediction
15. (C) Greedy decoding always chooses the word with the highest probability, whereas top-k sampling selects from the top-k highest probability words

References

1. Saed Rezayi, Zhengliang Liu, Zihao Wu, Chandra Dhakal, Bao Ge, Haixing Dai, Gengchen Mai, Ninghao Liu, Chen Zhen, Tianming Liu, et al. Exploring new frontiers in agricultural nlp: Investigating the potential of large language models for food applications. *IEEE Transactions on Big Data*, 2024.
2. Daniela Mechkaroska, Ervin Domazet, Amra Feta, and Ustijana Rechkoska Shikoska. Architectural scalability of conversational chatbot: the case of chatgpt. In *Future of Information and Communication Conference*, pages 54–71. Springer, 2024.
3. A Vaswani. Attention is all you need. *Advances in Neural Information Processing Systems*, 2017.
4. Amir Hossein Abaskohi. Navigating transformers: A comprehensive exploration of encoder-only and decoder-only models, right shift, and beyond. https://medium.com/@amirhossein.abaskohi/navigating-transformers-a-comprehensive-exploration-of-encoder-only-and-decoder-only-models-right-a0b46bdf6abe, 2023. Accessed: 2024-10-05.

5. Hannah Szmurlo and Zahid Akhtar. Digital sentinels and antagonists: The dual nature of chatbots in cybersecurity. *Information*, 15(8):443, 2024.
6. Victor Sanh, Lysandre Debut, Julien Chaumond, and Thomas Wolf. Distilbert: A distilled version of bert. *arXiv preprint* arXiv:1910.01108, 2019.
7. Jared Kaplan, Sam McCandlish, Tom Henighan, et al. Scaling laws for neural language models. *arXiv preprint* arXiv:2001.08361, 2020.
8. Brian Lester, Rami Alabi, and Yao Liu. The power of scale for parameter-efficient prompt tuning. *arXiv preprint* arXiv:2104.08691, 2021.
9. Hjalmar Olsson and Mark Steedman. In-context learning and induction heads. *arXiv preprint* arXiv:2109.00813, 2021.
10. Alec Radford, Jeffrey Wu, Rewon Child, et al. Improving language understanding by generative pre-training. *arXiv preprint* arXiv:2205.01068, 2022.

Google Bard and Beyond 5

By the end of this chapter, you will learn about Google Bard and its functionality.

Google Bard can be considered a new revolution in the deployment of large language models (LLMs) [3]. Similar to Google AI, Bard is based on a highly trained algorithm with various datasets containing texts and codes [1]. Some of the areas in which it has been used are text generation, language translation, content creation, and answers to knowledge-based questions.

Google Bard leverages the framework of the Transformer, a type of neural network used for the processing of text sequences. Such architecture also allows Bard to identify and learn statistical dependencies between the words and phrases concerning numerous corpora.

Previous chapters gave a detailed description of the Transformer model, in which we discovered that it can learn long-range contextual dependencies and can produce syntactically as well as semantically valid text.

In this chapter, we will go deeper into how Google Bard expands the Transformer model. Key advancements include:

- **Larger Dataset Utilization**: Training on the larger and diverse text and code corpus also allows Bard to learn more about diverse relationships between the words and phrases and deepen their general knowledge and perform the tasks better.
- **Enhanced Neural Network**: Bard learns better relationships because of a more effective neural network, and this makes it a better performer across different tasks.
- **Advanced Attention Mechanism**: The attention mechanism in Bard allows the network to selectively focus on specific sections of the input sequence for particular tasks, enhancing the system's effectiveness in applications such as machine translation and question-answering.

We next turn to review what works well and what does not in the context of Google Bard's architecture and the identification of areas of application.

5.1 The Transformer Architecture

Google Bard and Claude 2 rely on the Transformer architecture that has claimed its place as one of the most significant developments in natural language processing [2]. More detailed analysis of the core parts of the architecture, such as self-attention mechanisms or position-wise feedforward networks, could be found in Chapter 2 where we introduce how these components changed the existing approach to the language processing tasks.

Google Bard extends the principles leveraged in the Transformer architecture and how it tries to capture relationships and dependencies in the context of the text [3]. It is thanks to this base that Bard is capable of creating responses, compositions, and other content that are not only creative but also relevant to the given context.

Chapter 2 offers a critical review of this architectural innovation and the underlying mechanics of the Transformer architecture for enhanced comprehension of its function and contribution to generative AI.

Incorporating these ideas, Google Bard has improved Transformer architecture. Being a chat-based second generation of PaLM integration, Bard applies the Lambda architectural framework to preprocess; postprocess; and even incorporates it into the text generation, language translation, creative content generation, and informative QA processes. The primary differences that one can define between the classic Transformer and Google Bard are as follows:

- **Dataset**: As is the case with most Transformer models, Google Bard is trained on comparatively condensed text data sources, and, as has been mentioned above, the model works with text and code data. Such a large set of information resources with doctors, including 1.56 trillion words, enables Bard to learn even finer detail on the relations between words and phrases than in the case of traditional Transformers based on datasets of several million words.
- **Neural Network**. The proposed Transformer model has an inherently limited number of parameters but the neural network in the prior art model is relatively smaller and contains only a few hundred million of parameters. In comparison, Bard uses a considerably more extensive network of 137 billion parameters. This increase in the Bard scale allows them to analyze more nuanced and subtle patterns in the data collected.
- **Attention Mechanism**: The original Generative Transformers employ a self-attention function that has a single attention head. Google Bard though employs a multi-headed attention which means it can pay attention to 12 different parts of the input text at once. This feature enhances Bard's capacity to handle and produce real contexts in much more detail.

Table 5.1 Key differences between transformer architecture and Google Bard architecture

Feature	Transformer architecture	Architecture of Google Bard
Dataset	Smaller dataset of text	Massive dataset of text and code
Neural network	Smaller neural network	More powerful neural network
Attention mechanism	Self-attention mechanism	Multi-head attention mechanism
Output	Text that is generally accurate and informative	Text that is more accurate, informative, and creative

- **Output**: Bear in mind that, traditional Transformers create text that is normally precise and informative in its way and Bard's improvements result in the creation of more innovative text that is precise and informative in its way. This is due to the fact that it has been trained with a greater number of data points as well as the utilization of a more complex neural network layout and attention mechanism.

In general, the architecture of Google Bard represents a significant advancement from the original Transformer model, demonstrating a superior ability to understand complex patterns and generate innovative, detailed information (Table 5.1).

5.2 Google Bard's Text and Code Fusion

Google Bard works with a huge and the most diverse set of materials including plain text and code from books, articles, websites, etc., and various code repositories. This extensive training enables Bard to comprehend relationships of statistics from one word and several phrases in different contexts.

The following is a list of data employed by Google Bard:

- **Books**: The examples that Bard gets exposed to span over novel, non-fiction, and textbooks in literature. This diversity beneficially expands the informational sphere which is familiar to Bard acquiring various kinds of both literary and factual materials.
- **Articles**: In fact, in the model numerous articles involve learning by means of news articles, blog posts, and articles published in scientific journals. From this exposure, Bard gets to learn different style of writers as well as as well as various topics topics of interest.
- **Websites**: Bard's training also includes a training dataset that comes from a number of websites which contains text such as product descriptions, social media interactions, and

actual forum discussions. This variety enables Bard to understand contextual differences in diverse forms of the online social space.
- **Code Repositories**: From this, Bard learns code repositories and thus its capability to reason with basic programming concepts such as names of variables and functions, and several keywords.

The wide and diverse database to draw from also improves the capacity of Bard to produce text that is accurate and informative. This particular large-scale training allows Bard to surpass other smaller language models trained in the large-scale. Nevertheless, self-supervised learning methods are pre-applied for Google Bard which in turn makes use of a dataset. More specifically, Bard relies on a technique known as masked language modeling, where the parts of the text are cloaked, and the model must guess the words. This method enables Bard to understand complex relations between words and enhances its over-arching focus to different parts of the text.

5.3 Strengths and Weaknesses of Google Bard

5.3.1 Strengths

- Accuracy and Informativeness: Google Bard excels at generating text that is both grammatically correct and factually accurate. It also produces creative and engaging content.
- Creativity: Bard demonstrates versatility in generating various types of text, including poems, code, and scripts. Its output can be both humorous and thought-provoking.
- Empathy: Bard is capable of understanding and responding to human emotions, creating text that is empathetic and compassionate.
- Continuous Learning: Bard constantly updates and refines its knowledge through ongoing training on extensive datasets, improving its capabilities over time.
- Accessibility: Bard is designed to be accessible to a wide range of users, regardless of age or ability.

5.3.2 Weaknesses

- Bias: The extensive datasets used for training may contain biases, which can result in Bard producing biased or discriminatory content.
- Misinformation: Bard has the potential to generate text that is factually incorrect or misleading, which can contribute to the spread of misinformation.
- Security: As a complex software, Bard may be vulnerable to security threats that could be exploited to create harmful or malicious content.

- Privacy: Bard collects and stores user data, which raises concerns about user tracking and targeted advertising.
- Interpretability: As a black-box model, Bard's internal workings are difficult to interpret, making it challenging to ensure the accuracy and fairness of its outputs.

5.4 Difference Between ChatGPT and Google Bard

Hence, even though ChatGPT and Bard are based on Transformer architecture, the implementation is different. While ChatGPT uses only a decoder-only network, Bard comes with both an encoder and decoder network.

GPT-4 and Bard are prompt chain LMs that are famous for their generating human-like text, language translation, and the creation of various forms of content and informative answers. However, they exhibit notable differences:

- **GPT-4**: OpenAI's GPT-4 is trained with a dataset, which has billions of words (exact numbers are unavailable). It is one of the largest LLMs to date and its strength is in the generation of creative text formats and patterns such as poems, codes, scripts, songs, emails, and letters. Like the previous versions, GPT-4 is very effective in generating responses to abundant questions, simple, complicated, general, complex, or even ambiguous.
- **Bard**: Developed by Google's artificial intelligence, Bard has been trained on an extensive corpus of up to 1.56 trillion words and currently operates with 137 billion parameters. This scale allows the model to integrate real-time data from Google Search, enhancing its ability to provide accurate and current responses. Bard excels in tasks that require problem-solving based on real-world knowledge, such as understanding humor and sarcasm, as well as in generating creative content. Its design enables it to process intricate patterns and generate contextually relevant information, positioning it as a powerful tool for applications in natural language understanding and content creation.

In other words, GPT-4 is overall better for tasks involving the need for greater depth of analysis and language comprehension such as translation and summarization. On the other hand, Bard is better equipped for the tasks that need to incorporate knowledge of the real world and access to extra information that does not belong to a knowledge base.

For further comparison, the following resources are included:

- "ChatGPT versus Bard": "Which Large Language Model Is Better?" by Jonathan Morgan (Medium).
- "ChatGPT versus Bard": "A Comparison of Two Leading Large Language Models" by Siddhant Sinha (Towards Data Science).
- "ChatGPT versus Bard": Google AI Blog's "Choosing Your Large Language Model."

- "ChatGPT versus Bard": Specifically, I compared the performance of the present state-of-the-art large language model or LLMs, taking the "Wave: A Performance Comparison" created by the PaLM Team to reflect Google AI.
- "ChatGPT versus Bard": Currently, there are papers such as "A Bias Comparison" written by the AI Ethics Team comprising of employees from Google AI.

All these sources offer a clear comparison between ChatGPT and Bard that includes features, potential drawbacks, and even performance. They also take care of the possible bias in each of the models. Just to be aware these evaluations are not set to be final since both models are still undergoing the process of development and enhancement.

5.5 Claude 2: A Plan For A Qualitative Connection Between Human and Computers

In the last 10 years, technological advancement has seen machines incorporated with artificial intelligence thus being given remarkable functions. However, the analyzed results and comparisons with reference data indicate that despite the presence of correlations there exists a phenomenal difference between human intelligence and machine cognition. Although there are very effective AI systems designed for a specific purpose, designing an AI system that understands tacit knowledge, conversing in context, or emulating human reasoning is not an easy task.

Claude, which has been developed by the company Anthropic, can be regarded as an example that effectively bridges this gap. Safety first, benevolence as well as integrity are the principles that Claude was aimed to follow, to become an example of the next step toward human-oriented AI. In integrating natural language processing with strong human-centered concepts, Claude makes the application of artificial intelligence very natural, straightforward, and humane.

5.5.1 Key Features of Claude 2

Here are some of the standout features that set Claude 2 apart from other chatbots:

- Multiturn Conversational Ability: While chatting with Claude 2 it is also possible to observe how the program interacts by choosing the subsequent lines of the conversation and does not stop at one or two dialogs, but sees the discussion as a non-intersecting process.
- Enhanced Reasoning: In comparison with Claude 1, Claude 2 shows a higher level of performance in the logical connections and inferences made referring to the current conversational context.

5.5 Claude 2: A Plan For A Qualitative Connection Between Human and Computers

- More Natural Language: Claude 2 is designed to sound slightly less "mechanical" and more "natural"; it's written in plain language that is free of contrivance.
- Broad Conversational Range: Complexity: Claude 2 can hold a virtually unlimited number of talks on various issues of every day conversational topics such as sports, movies, and music.
- Customizable Personality: Claude 2 is able to provide different personas for each character which has different attributes such as focused, balanced, or playful and users have the chance to select a certain style they like.
- Feedback System: The user can give feedback on Claude 2 responses the outcome of which is then utilized to calibrate Claude 2 for increased performance in the future.

Claude 2 presents the possibility of AI development becoming more natural following the rationality of the human mind and diminishing the gap between artificial intelligence and humans.

5.5.2 Comparing Claude 2 to Other AI Chatbots

Applicants such as Claude 2 are relatively young and can become a worthy counterpart to the most advanced systems identified at the moment, including Google's LaMDA and Microsoft's Sydney. A model employing Microsoft Sydney, a chatbot deployed in Bing since the end of 2020, was preliminarily trained on models earlier used in India. Also, it is a large language model (LLM) that can generate text, translate languages, and provide an informative answer as ChatGPT and Bard. Here's how Claude 2 stands out from its competitors:

- Enhanced Conversational Intelligence: Of Claude 2 and Microsoft's Sydney, the former is shown to have better conversational reasoning as it provides better advice and relies on contextual information when conversing.
- Different Focus Than LaMDA: However, Google's LaMDA is highly creative in comparison to Claude 2 which creates logically based solutions and is better for users who want an accurate logical approach.
- Broader Availability: Claude 2 is more restricted compared to both LaMDA and Sydney, but unlike those two models, Anthropic has announced that Claude 2 will be made publicly available sooner rather than later.
- Ethical Considerations: Claude 2 does not venture into the controversy around LaMDA as it has been pulled into the ethical questions regarding sentence. Anthropic draws attention to the fact that Claude 2 has no inner life per se and only exhibits deterministic decision-making based on coherent human values.
- Open Feedback System: Claude 2 also engages users explicitly to provide feedback and feedback from the user unlike in LaMDA or Sydney. Openness of information

also encourages the constant improvement of the website and extremely fast constant tweaking.

Using these features, Claude 2 positions itself as a worthy competitor for other AI chatbots available on the market that provide reasoning and a convenient interface for users while engaging them with a truly unique experience.

5.5.3 The Human-Centered Design Philosophy of Claude

Hence, according to Claude's design principles, its interactions with humans are to be positive, clear, and in compliance with ethical norms. Here are the key guiding principles:

- Helpful over Harmful: Claude's primary mission is to assist users while rigorously avoiding actions that could cause harm.
- Honest over Deceptive: Built on a foundation of truthfulness, Claude is designed to communicate honestly, even when uncertain, ensuring it does not mislead users.
- Transparent over Opaque: Claude emphasizes transparency by clearly explaining its decision-making processes and capabilities when prompted, fostering trust and openness with users.
- Empowering over Exploitative: Claude aims to empower users with valuable information, consciously avoiding any exploitation of human vulnerabilities for personal or commercial gain.
- Collaborative over Competitive: As a supportive partner, Claude enhances human capabilities rather than competing with or replacing them, focusing on collaboration.
- Ethical over Unethical: Claude's behavior is guided by ethical principles embedded in its training, ensuring it aligns with human values and promotes virtuous interactions.

These are the principles on which Claude was developed with a clear focus on humans to ensure that there is a good relationship between an AI assistant and the users.

5.5.4 An Analysis of AI Conversational Competencies of Claude

Claude is intended to be an AI equipped with human-like experience as its core built on top of natural language processing technologies. Key features include:

- **Large Language Models**: Claude leverages Transformer-based neural networks, similar to GPT-3 and LaMDA, enabling it to understand and respond to the subtleties of human language.

- **Reinforcement Learning through Feedback**: Claude improves its responses through continuous learning, guided by interactive human feedback, which refines its performance over time.
- **Commonsense Reasoning**: Claude's extensive training allows it to make informed inferences about concepts it hasn't been explicitly trained on, enabling more intuitive interactions.
- **Constitutional AI Safeguards**: Claude operates within ethical boundaries designed to prevent it from engaging in harmful, unethical, or illegal activities.
- **Internet-Scale Self-Supervised Learning**: Claude continuously expands its knowledge base by processing large amounts of unstructured public data from the Internet.
- **Natural Conversation Flow**: Claude excels at managing multiturn, open-ended dialogs, facilitating smooth and authentic conversational exchanges.

These proficiencies let Claude provide a near-human-like conversational experience that is unadulterated by the ill effects of over-automation or a business-like approach.

5.6 What is Constitutional AI?

To eliminate the shortcomings of conventional techniques, Constitutional AI applies feedback by applying a set of guidelines together with AI to analyze outcomes. The term "Constitutional" reflects the model's adherence to these guiding principles, which dictate its behavior. At its core, the constitution steers AI toward normative behaviors aligned with these principles, such as avoiding toxic or discriminatory content, refusing to assist in illegal or unethical activities, and promoting an overall system that is helpful, honest, and harmless.

Claude 2 employs a constitutional AI framework, where a set of guiding principles shapes its training and behavior, ensuring it avoids generating harmful or offensive content. This concept is illustrated in Fig. 5.1, drawing from the research published by Yuntao Bai and colleagues at Anthropic.

As Fig. 5.2 indicates, the constitution significantly functions in two dynamics of Claude's maturation: The first is in Claude's process of exiting childhood, whereby the constitution continues reading → During the first phase of development the model is programmed with "rules of engagement" and tested and refined using example scenarios. In the second stage, the model gets into reinforcement learning again but this time it learns from AI feedback that is strictly constitutional. Such an approach helps Claude 2 to advance or, at least, not to regress while guaranteeing compliance of its outputs with the principle of maxims' harmlessness.

- Nonmaleficence: Claude 2 should avoid causing harm to individuals or society.
- Beneficence: Claude 2 should act in ways that benefit people and society.

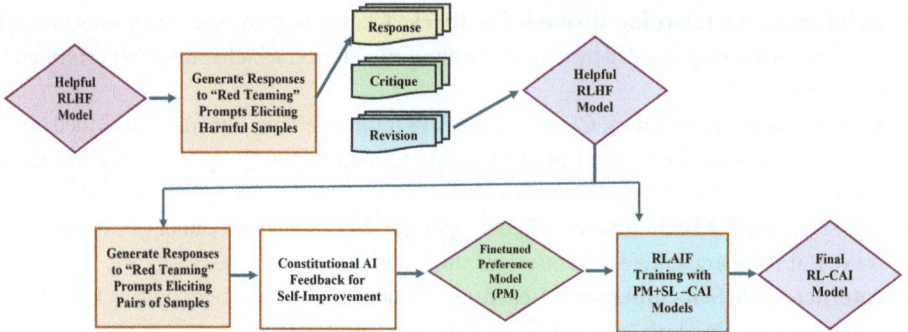

Fig. 5.1 Constitutional AI Process: The basic steps of the Constitutional AI (CAI) process involve both a supervised learning (SL) stage, shown at the top, and a reinforcement learning (RL) stage, depicted at the bottom of the figure. In both stages, critiques and AI feedback are guided by a small set of principles derived from a "constitution." The supervised stage is much beneficial to amplify the preliminary model for managing and configuring behavior at the beginning of the RL phase and for dealing with exploration issues. The RL stage lets the plan then increase performance and reliability in the next RL stage (reproduced from [4])

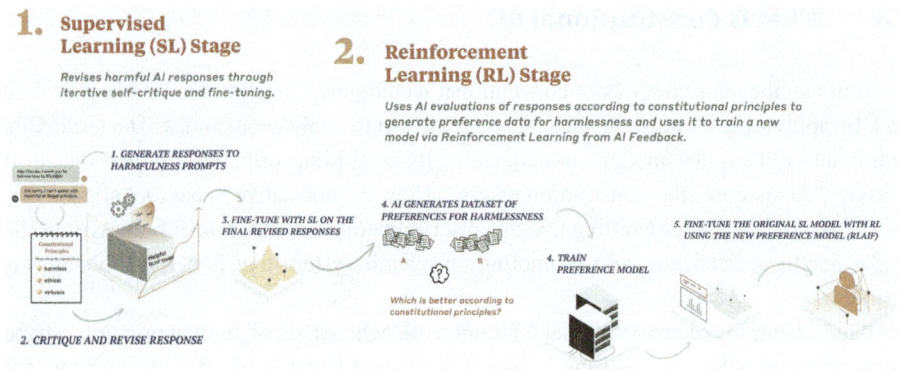

Fig. 5.2 Claude's constitution by anthropic [4]

- Justice: Claude 2 should treat all individuals fairly and equitably.
- Autonomy: Claude 2 should respect human autonomy.
- Privacy: Claude 2 should safeguard human privacy.
- Accountability: Claude 2 should be responsible for its actions.

These principles are embedded in Claude 2's development in several ways. First, they regulate the process of selecting training data, so that this content violates these principles. Second, the principles are used to measure the model's quality and from the outcome, penalties are given when the model produces outputs that are contrary to the principles.

Hence, the feedback loop assists Claude 2 in maintaining an effective way of avoiding any dangerous or abusive results continually.

Constitutional AI is another approach that seeks to level the incorporation of AI systems with human values as well as ethical principles; nevertheless, it is not free from limitations. AI systems are always complicated, however, the undesirable output that can be potentially created is still a possibility no matter the precautions taken. Hence, there is a need to propose better measures like stringent safety measures to avoid mishandling or wrong use of advanced technology like AI.

5.7 Claude 2 Versus GPT-3.5

Claude 2 and GPT-3.5 are both neural language generation models that can write text, translate languages, or answer questions. Still, there are some differences between the given models:

- **Training Data**: Claude 2 is built on a diverse dataset, which includes both texts and code and therefore more versatile and produces more accurate and detailed answers. In contrast, GPT-3.5 is developed solely on the basis of textual data.
- **Safety Features**: Claude 2 contains additional precautions intended to avoid the creation of any forms of objectionable/reprehensible/inaccurate content involving diverse filters and detection algorithms for handling biased outputs. GPT-3.5 lacking these general security measures is more vulnerable to creating hazardous or even obscene content (Table 5.2).

Claude 2's design focuses on integrating traits like common sense, conversational fluency, and alignment with human values, marking significant progress in bridging the gap between human and machine intelligence. Instead of diminishing human capacities, Claude's human-

Table 5.2 Comparison between claude 2 and GPT-3.5

Feature	Claude 2	GPT-3.5
Training data	Text and code	Text only
Safety features	Yes	No
Target audience	Businesses, governments, individuals	Entertainment
Accuracy	More accurate	Less accurate
Safety	Safer	Less safe
Versatility	More versatile	Less versatile

oriented design and the natural language skills in the collaboration draw the vision of the humans and AI interaction closer to reality.

5.8 Other Large Language Models

Beyond ChatGPT, Google Bard, and Claude, several other large language models (LLMs) are being developed, each with unique capabilities. These models are trained on vast datasets, enabling them to perform tasks such as text generation, translation, question-answering, and code generation.

5.8.1 Falcon AI

Falcon AI, developed by the Technology Innovation Institute (TII) in the United Arab Emirates, is a notable LLM. Falcon AI can perform a wide array of tasks, including:

- **Text Generation**: Falcon AI can generate text, translate languages, create various forms of content, and provide informative answers.
- **Natural Language Understanding**: It comprehends text meaning and responds comprehensively and informatively.
- **Question-Answering**: Falcon AI can handle open-ended, complex, or unusual questions with accurate responses.
- **Summarization**: It can distill long texts into concise summaries.
- **Code Generation**: Falcon AI can generate code in languages such as Python or Java among other programming languages.
- **Data Analysis**: Contained in this repository, it can analyze given datasets and extract insights from them.

However, even in its developmental phase, Falcon AI demonstrates high potential in a variety of contexts and can be particularly useful in solving problems that involve text data analysis for decision-making; at the same time, users should be aware of the intrinsic bias of large language models.

- **Falcon AI Models**
  Falcon AI has two models:

  – Falcon 180B: A 180-billion parameter model adept at handling complex tasks like language translation, creative content generation, and in-depth question-answering.
  – Falcon 40B: A more efficient 40-billion parameter model optimized for tasks requiring less computational power.

5.8 Other Large Language Models

- **Notable Applications**
  Falcon AI is already being utilized in various fields:

  - PreciseAG: Provides insights into poultry health.
  - DocNovus: Allows users to interact with business documents, receiving responses as if consulting with an expert.
  - Healthcare, Education, and Finance: Falcon AI is being explored in these industries for developing advanced applications.

- **Key Features and Limitations**

  - **Key Features**:

    - Falcon AI is a 180-billion parameter autoregressive model which allows the generation of text but does not possess deep cognition.
    - A general training as it trains on voluminous text and code databases, hence its knowledge and skills are general.

  - **Limitations**:

    - As a large language model, it is susceptible to biases.
    - Being under development, it may struggle with certain tasks.
    - Responsible usage is essential, given these limitations.

Due to its impressive performance as a language model, Falcon AI still has several potential as technology users are still advised to adopt various norms that the technology has that are not good depending on how the technology develops.

5.8.2 LLaMa 2

The second one is LLaMa 2, a set of large language models of Meta AI, the next generation of LLaMa models released in July 2023.

- **Key Improvements**:

  - Expanded Training Data: The newer version of LLaMa is claimed to have been trained on a dataset of two octillion tokens of text and code that is twice the dataset

used in the initial LLaMa. This is so as the data implies the model gains new data and with the implication the model's knowledge is extended.
- Extended Context Length: The new model allows for a longer context thereby enhancing its capability of understanding and generating text over longer passages, which is significant in question-answering as well as summarization.

- **Architectural Modifications**:

  - **Pre-Normalization**: Detailed in the LLaMa 2 paper as modularity 2, this new version contains an additional pre-normalization feature that normalizes inputs before passing through each layer. This adjustment makes the model more stable and optimize it for its best performance.
  - **SwiGLU Activation Function**: The activation function used in the current SwiGLU model is different from that used in the previous models; the ReLU function has been replaced in the current SwiGLU activation function. Compared with other existing solutions, such as SwiGLU, it is more efficient and effective and leads to higher model performance.
  - **Rotary Positional Embeddings**: Repeatedly, LLaMa 2 does not apply sinusoidal embeddings as is done in LLaMa 1 but instead applies rotary positional embeddings. This method offers a better way of encoding positional information within the text when compared to the traditional approaches.

- **Additional Features**:

  - **Larger Context Window**: The model processes larger amounts of information at once, thanks to its expanded context window.
  - **Grouped-Query Attention**: This feature optimizes the model's attention mechanism, improving how it manages and responds to input text.

Like its predecessor, the LLaMa 2 follows the workings of the Transformer model built with encoder and decoder layers to map the input text into the hidden representation and vice versa for the output text.

- **Performance and Applications**:
As it was reported, the new LLaMa 2 sustains the game over the original LLaMa in text generation, translation, some question-answering, and code generation. Said to be more helpful and safe by reinforcement learning from human feedback (RLHF), it is used.
In the present day used in conversation models, code production, and question analyzing, LLaMa 2 could be used further, particularly in areas like education, healthcare, and customer care.

5.8 Other Large Language Models

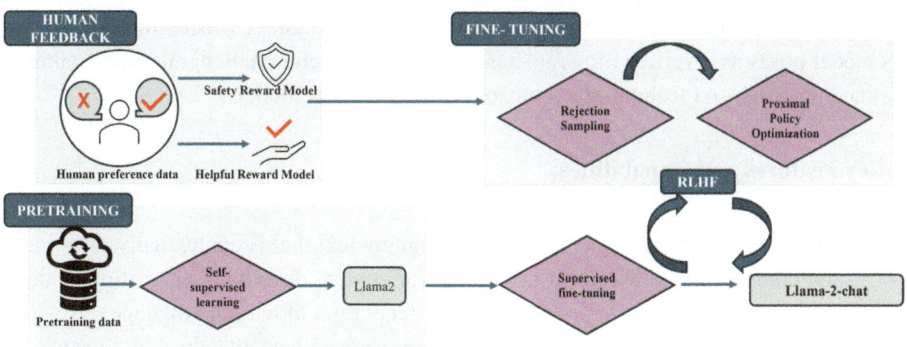

Fig. 5.3 Training process: Helps to explain what kind of training LLaMa 2-Chat went through. The process starts with pre-training at open sources on the Internet. After this, there is the first version that is generated of LLaMa 2-Chat by supervised fine-tuning. The model is then improved using reinforcement learning with human feedback methodologies such as rejection sampling and proximal policy optimization (PPO). In the RLHF stage, data corresponding to reward modeling as well as improvements to the model are collected to make sure the reward models are up to date with the distribution. (reproduced from [5])

- **Key Features**:

    - Its training has been done on a large-scale textual as well as code information set.
    - Has fifty billion tokens, however contains two trillion tokens.
    - Higher context length than the original LLaMa.
    - Includes concepts such as grouped-query attention as fresh additions.
    - It is alright in several areas as evidenced by higher performances on the benchmarks.
    - There is better helpfulness and safety with the use of RLHF.

- **Limitations**:
  Systematic bias is simply built into the use of large language models.
  In general, I can note that LLaMa 2 is a solid language model that can be used for numerous purposes. However, some precautions are still required together with recognition of its drawbacks in order to gain the greatest benefits from its implementation (Fig. 5.3).

5.8.3 Dolly 2

Dolly 2 is the latest and largest model created by Databricks with 175 billion parameters which is a causal language model. In this regard, expanding upon the features of the previous models, Dolly 2 works with a large and rather heterogeneous dataset, from texts and code, making it a truly universal tool in the context of natural language processing. As a result, this model poses as a variety of parameters to work on several challenging tasks making it a refined growth in AI technology compared to the previous models.

- **Key Features and Capabilities**:

  - **Text Generation**: Dolly 2 is good at designing text that is contextually meaningful and contextually coherent. It is suitable for content development, writing of documents, or any other task that requires the generation of natural language.
  - **Translation**: The model can also involve translation, this will help to bring handicap-free ways of multilingual communication.
  - **Question-Answering**: Dolly 2 comes with a great ability to answer questions appropriately and in detail. It is also capable of responding to both absolute and question-formatted types of queries, thus good for finding information and customer service.
  - **Code Generation**: Servo-wired to create, and support the coding need, Dolly 2 possesses the ability to create code snippets in commonly used programming languages to support software production as well as automated coding.
  - **Data Analysis**: It can also work on the data and make decisions, distill knowledge, and condense information within the shortest period.
  - **Summarization**: Dolly 2 is equipped with the ability to read whole articles and extract the most important information in order to help a user focus on a particular passage for a shorter period of time.
  - **Creative Writing**: It can create articles, stories, essays, and poems as creative content to exemplify its use in the creative industries.

- **Technical Specifications**:

  - **Parameter Count**: Dolly 2 is equipped with 175 billion parameters or 103 billion more than in the previous models. This large parameter count also further improves the model's capacity to comprehend and synthesize intricate linguistic features.
  - **Training Data**: It is important to note that the training dataset used in the model is all-inclusive and also encompasses several text and code sources. This kind of diversity in the training set is important because Dolly 2 is able to learn from all types of content and perform well in various tasks.
  - **Development Status**: Currently there is Dolly 2 in production. Application of the model has shown its efficiency, and further improvements are expected to broaden its effectiveness (Table 5.3).

5.8 Other Large Language Models

Table 5.3 Comparison of Google Bard, claude 2, GPT-3.5, falcon AI, and LLaMa 2

Feature	Google Bard	Claude 2	GPT-3.5	Falcon AI	LLaMa 2
Developer	Google AI	Anthropic	OpenAI	Technology Innovation Institute	Meta AI
Architecture	Transformer with multi-head attention	Transformer-based with additional innovations	Transformer (decoder-only)	Autoregressive decoder-only	Transformer-based with enhancements
Training data	1.56 trillion words (text and code)	Diverse text and code datasets	Text data (size not disclosed)	1 trillion tokens (text and code)	2 trillion tokens (text and code)
Parameters	137 billion	Not specified	Not specified	180 billion (Falcon 180B)	Not specified
Attention mechanism	Multi-head attention	Advanced multi-head attention	Single attention head	Not specified	Not specified
Text generation	Yes	Yes	Yes	Yes	Yes
Language translation	Yes	Yes	Yes	Yes	Yes
Code generation	Yes	Not specified	Not specified	Yes	Not specified
Summarization	Yes	Not specified	Yes	Yes	Yes
Question-answering	Yes	Yes	Yes	Yes	Yes
Safety features	Not specified	Advanced safety mechanisms	Basic safety features	Bias filters, detection systems	Not specified
Creativity	High	High	High	Moderate	Moderate
Versatility	High	High	High	High	High
Privacy concerns	Data collection	Emphasis on user feedback	Less emphasis on privacy features	Not specified	Not specified
Bias and misinformation	Potential for bias and misinformation	Ethical principles to minimize bias	Potential for bias and misinformation	Potential for bias	Potential for bias

- **Applications and Potential**:

    - **Dialog Systems**: Dolly 2 is used in conversational agents and chatbots, providing users with engaging and contextually appropriate interactions.
    - **Code Assistance**: It supports developers by generating and refining code, streamlining the coding process, and reducing manual effort.
    - **Customer Support**: The model's question-answering capabilities are employed in customer service applications, offering accurate and helpful responses to user inquiries.

- **Content Creation**: Dolly 2's text generation and creative writing abilities make it a valuable tool for content creation, including marketing materials, articles, and more.

The emerging capability of Dolly 2 makes it possible to become an important technology in diverse fields such as education, health, and even business practices. Its continuous improvement is envisioned for the purpose of improving its functions and fixing its flaws so that its uses can be beneficial and safe in different fields.

From the case of Dolly 2, we get a sense of a vast improvement in language modeling with enhanced features and flexibility for various uses. The number of parameters and amount of training data are larger than that of other similar models, which makes this model more versatile and efficient. Due to the several applications that Dolly 2 has the potential to be used on, it is crucial in the future of AI-driven solutions even if it is still under development.

The table below compares large language models discussed in the lecture, namely, Google Bard, Claude 2, and GPT-3.5, Falcon AI, and LLaMa 2.

5.9 Summary

There are currently many LLMs, with several of the most notable ones being ChatGPT, Google Bard, Claude, and several LLMs, others among them being Falcon AI, LLaMa 2, and Dolly 2. Known as hallmark models of deep learning, these advanced models are learned on huge volumes of text and code and aspire to the following abilities within text generation, translation, question-answering, and code generation. On this basis, as a technology that sets the basis for the LLMs, this technique is in a firm position to provide new solutions to many different applications in the future. This ability to process and generate human-like text makes them useful tools in application areas such as dialog systems, content generation, or code generation. Thus, it is highly imperative to fight different biases but to acknowledge the weaknesses and the ethics involved in pursuing an LLM. There are problem areas that include bias and self-organizing capacity, which might create risks of misuse. It is thus vital that the technologies are employed responsibly and appropriately to foster their impact on society and in the right ethical way possible. The further development of LLMs is not even expected to offer more sophisticated and versatile models for the future because the development of artificial intelligence is a continuous process and it is proving its effect more and more in each and every sector.

5.10 Multiple-choice Questions

In this section, you'll find a series of multiple-choice questions designed to test your understanding of key concepts in generative AI. Choose the correct answer for each question.

1. What is a key feature of Google Bard's architecture?

    (A) Decoder-only structure
    (B) Transformer architecture
    (C) Recurrent Neural Network (RNN) architecture
    (D) Autoencoder architecture

2. How does Google Bard's dataset size compare to traditional Transformer models?

    (A) Bard uses a smaller dataset than traditional Transformers
    (B) Bard uses a dataset that includes both text and code, significantly larger than traditional datasets
    (C) Bard's dataset is identical in size to traditional Transformer models
    (D) Bard uses only a small subset of traditional Transformer datasets

3. What type of attention mechanism does Google Bard use?

    (A) Single-head attention
    (B) Multi-head attention
    (C) Self-attention with feedforward networks
    (D) Cross-attention

4. Which component of Google Bard's architecture is significantly larger than traditional Transformer models?

    (A) Training data
    (B) Neural network parameters
    (C) Context length
    (D) Training time

5. What is one key difference between ChatGPT and Google Bard?

    (A) ChatGPT uses an encoder–decoder architecture while Bard uses a decoder-only architecture
    (B) Bard integrates information from Google Search, while ChatGPT does not
    (C) Bard was developed by OpenAI, whereas ChatGPT was developed by Google AI
    (D) ChatGPT uses a multi-head attention mechanism, while Bard uses a single-head attention mechanism

6. What type of data does Google Bard's training dataset include?

    (A) Only text
    (B) Only code
    (C) Text and code
    (D) Only images

7. Which principle does Claude 2 NOT adhere to?

    (A) Honesty
    (B) Transparency
    (C) Exploitative behavior
    (D) Beneficence

8. What is a key feature of Claude 2's conversational ability?

    (A) Limited conversational range
    (B) Enhanced logical reasoning
    (C) Rigid language style
    (D) Lack of customization options

9. What distinguishes Falcon AI from other language models?

    (A) Its use of images in training data
    (B) Its 180-billion parameter autoregressive model
    (C) Its focus on creative content generation
    (D) Its implementation of reinforcement learning

10. What feature does LLaMa 2 improve upon compared to its predecessor?

    (A) Expanded context length
    (B) Reduced parameter size
    (C) Lack of text data
    (D) No improvements over LLaMa

11. Which of the following is a strength of Google Bard?

    (A) Generating text that is generally accurate and informative
    (B) Producing highly biased and discriminatory content
    (C) Lack of creativity in text generation
    (D) Inability to understand human emotions

12. Which AI model emphasizes a human-centered design philosophy?

    (A) GPT-3.5
    (B) Claude 2
    (C) Falcon AI
    (D) LLaMa 2

13. What is one application of Falcon AI?

    (A) Social media analysis
    (B) Text summarization
    (C) Image recognition
    (D) Music composition

14. Claude 2's constitutional AI framework aims to prevent:

    (A) Creative content generation
    (B) Harmful, unethical, or illegal activities
    (C) Accurate and informative responses
    (D) Continuous learning and improvement

15. What is a key limitation of GPT-3.5 compared to Claude 2?

    (A) It is trained on both text and code
    (B) It includes advanced safety features
    (C) It lacks comprehensive safety mechanisms
    (D) It has more parameters than Claude 2

5.11 Answers

Below are the answers to the multiple-choice questions from the previous section:

1. **(B)** Transformer architecture
2. **(B)** Bard uses a dataset that includes both text and code, significantly larger than traditional datasets
3. **(B)** Multi-head attention
4. **(B)** Neural network parameters
5. **(B)** Bard integrates information from Google Search, while ChatGPT does not
6. **(C)** Text and code
7. **(C)** Exploitative behavior
8. **(B)** Enhanced logical reasoning

9. **(B)** Its 180-billion parameter autoregressive model
10. **(A)** Expanded context length
11. **(A)** Generating text that is generally accurate and informative
12. **(B)** Claude 2
13. **(B)** Text summarization
14. **(B)** Harmful, unethical, or illegal activities
15. **(C)** It lacks comprehensive safety mechanisms

References

1. Imtiaz Ahmed, Mashrafi Kajol, Uzma Hasan, Partha Protim Datta, Ayon Roy, and Md Rokonuzzaman Reza. Chatgpt versus bard: A comparative study. *Engineering Reports*, page e12890, 2024.
2. Saleh Obaidoon and Haiping Wei. Chatgpt, bard, bing chat, and claude generate feedback for chinese as foreign language writing: A comparative case study. *Future in Educational Research*, 2024.
3. Mohaimenul Azam Khan Raiaan, Md Saddam Hossain Mukta, Kaniz Fatema, Nur Mohammad Fahad, Sadman Sakib, Most Marufatul Jannat Mim, Jubaer Ahmad, Mohammed Eunus Ali, and Sami Azam. A review on large language models: Architectures, applications, taxonomies, open issues and challenges. *IEEE Access*, 2024.
4. Anthropic. Claude's constitution. https://www.anthropic.com/news/claudes-constitution, 2024.
5. Hugo Touvron, Louis Martin, Kevin Stone, Peter Albert, Amjad Almahairi, Yasmine Babaei, Nikolay Bashlykov, Soumya Batra, Prajjwal Bhargava, Shruti Bhosale, et al. Llama 2: Open foundation and fine-tuned chat models. *arXiv preprint* arXiv:2307.09288, 2023.

Diffusion Model and Generative AI for Images   6

By the end of this chapter, you will:

- **Understand the fundamentals of Variational Autoencoders (VAEs)**: Learn the core concepts of VAEs, including their architecture and how they are used for generative tasks.
- **Gain insight into Generative Adversarial Networks (GANs)**: Explore the principles behind GANs, including their structure, training process, and applications in generating high-quality images.
- **Explore diffusion models and their types**: Understand the workings of diffusion models, including their architecture and different types, and their role in generating realistic images.
- **Comprehend the technology behind DALL-E 2**: Understand the CLIP training process used in DALL-E 2 and the specifics of the text-to-image generation process, understanding how these technologies combine to produce creative and accurate images from textual descriptions.
- **Learn about stable diffusion and the Latent Diffusion Model (LDM)**: Learn what Stable Diffusion fundamentally is and discusses the efficiency of high-quality image generation through the latent diffusion model.
- **Learn the technology behind midjourney**: Evaluate the generative techniques and methodologies that Midjourney uses in developing images from text descriptions.
- **Differentiate between DALL-E 2, stable diffusion, and Midjourney**: Categorize these latest text-to-image models and discuss some of their attributes, utility, as well as some of their differences.
- **Understand the role of data augmentation and preprocessing**: Find out how data augmentation and preprocessing enhance predictive models' performance and increase the quality of the derived results.

© The Author(s), under exclusive license to Springer Nature Switzerland AG 2026
D. Bhati et al., *A Beginner's Guide to Generative AI*, Synthesis Lectures on Computer Science, https://doi.org/10.1007/978-3-031-84724-0_6

- **Explore the use of attention mechanisms in image generation**: Get insight into how attention mechanisms improve the generation process and the efficiency of text-to-image models.
- **Grasp the key loss functions and optimization techniques**: Learn about the loss functions and the possibilities for optimization, which are used in training GANs and other generative models to get the best results.
- **Identify the benefits and applications of generative models**: Explain the real-life use of generative models with respect to the different domains and the way emerging fields like art, design, and entertainment are being influenced by generative models..

Generative models are widely used in machine learning and the major advancements in the creation of new data samples are accomplished through them by using Generative Adversarial Networks (GANs), Variational Autoencoders (VAEs), and other models [1]. This chapter will begin with the clarification of the understanding of the models that are under discussion, and only then will we proceed to a detailed explanation of the mechanics of diffusion models.

- **Generative Adversarial Networks (GANs)**: GANs refer to a competition that exists between two neural networks [2]. Another network, named the generator, attempts to generate data samples that are real, while the other named, the discriminator, attempts to classify between real data and fake data produced by the generator. In each iteration, the generator enhances the model's capability to generate credible data, on the other hand, the discriminator enhances its capability to identify fake data. Since this back-and-forth process is used to generate reasonable numbers of images, perceptible videos, and other datasets, GANs provide highly realistic results. However, training GANs can be challenging because they can suffer from issues like mode collapse (where the generator produces limited varieties of outputs) and gradient vanishing (where the model stops learning effectively) [3].
- **Variational Autoencoders (VAEs):** VAEs work differently by compressing data into a compact, latent space and then reconstructing it [4]. Imagine you have a complex image, and you want to describe it in a simplified way. The encoder compresses this image into a smaller, more manageable form, and the decoder reconstructs it from this simplified version. VAEs are built on a solid theoretical foundation but can struggle with generating very sharp or detailed outputs due to challenges in defining effective loss functions.
- **Diffusion Models**: Diffusion models offer another approach. Imagine you have a picture, and you start adding random noise to it until it's completely unrecognizable [5]. A diffusion model becomes aware of how to undo this process in a step-by-step manner to generate the original image. The model learns how the data turns into the noise over time and in its turn, it attempts to reverse this process to the data. Unlike VAEs which learn a probabilistic reconstruction of the data, diffusion models are trained on the understanding

of the manner in which noise affects the data, and also for the same reason, they are able to generate the data.

However, to understand the features of the diffusion models in detail, it is necessary to consider the GANs and VAEs' fundamental concepts. Both methods also have their advantages and limitations, and analyzing them provides a good foundation for understanding how the use of diffusion models enhances these strategies.

6.1 Understand the Fundamentals of Variational Autoencoders (VAEs)

Variational Autoencoders (VAEs) are generative models that are based on the Bayesian approach for the combination of prior knowledge in the form of probabilistic models with deep neural networks for learning and generating data distributions [1]. Here's a brief understanding of what each of them is all about, the manner they're constituted, and what they are capable of:

6.1.1 Core Concepts of VAEs

- **Latent Variables**:

  - **What They Are**: The second type of variable requires more explanation as it is called a latent variable, and it is an abstract characteristic that characterizes the data. For instance, let us consider a dataset containing face images; latent variables may involve hidden and intrinsic properties such as age or hair color, which are normally not easily discernible.
  - **How They Work**: VAEs employ the observed data x to model it as coming from these latent variables z. The conditional probability with which the observed data x can be obtained from the latent variables z is given by $p(x \mid z)$. Based on this model, the decoding network reconstructs x from z.

- **Approximate Posterior**:

  - **What It Is**: In order to compute the latent variables z from the data-generating process x, VAEs employ an approximate posterior distribution of $p(z \mid x)$. This distribution is learned by the encoder network, which transforms the data x to the latent space.

- **Evidence Lower Bound (ELBO)**:

    - **What It Does**: Directly maximizing the likelihood of the data is often complex. Instead, VAEs optimize a simpler measure called the Evidence Lower Bound (ELBO), defined as

    $$\text{ELBO} = \mathbb{E}_{q(z|x)}[\log p(x \mid z)] - \text{KL}[q(z \mid x) \| p(z)]$$

    - The first term, the **Reconstruction Term** $\mathbb{E}_{q(z|x)}[\log p(x \mid z)]$, measures how well the model can reconstruct the data from the latent variables.
    - The second term, the **KL Divergence** $\text{KL}[q(z \mid x) \| p(z)]$, ensures the latent space distribution is regularized and does not deviate significantly from the prior distribution $p(z)$.

6.1.2 Architecture of VAEs

- **Encoder**:

    - **Role**: The encoder compresses the input data into a latent space. It produces parameters (like the mean and variance) that describe a distribution from which the hidden features are sampled.
    - **How It Works**: Imagine you're simplifying a detailed image into a smaller, abstract representation. This is what the encoder does using a neural network.

- **Decoder**:

    - **Role**: The decoder takes these abstract representations (latent variables) and tries to reconstruct the original data from them.
    - **How It Works**: If the encoder is like a summarizer, the decoder is like a re-creator.

6.1.3 Generative Tasks with VAEs

- **Sampling**: VAEs can come up with new examples since they learn another distribution in the latent space. For example, the VAE you trained on faces will allow you to generate new faces by drawing samples from the latent space and apply a decoder.
- **Interpolation**: VAEs make it possible for you to mix several characteristics of two data points. For instance, you can generate new faces in between two people's latent variables and end up with faces that have the features of the two persons.

- **Data Reconstruction**: As noted earlier, VAEs can deconstruct data from the abstract representations that encodes the data. This capability is useful in applications such as denoising (where noise has to be removed) and data generation that has been patternized.

6.2 Gain Insight Into Generative Adversarial Networks (GANs)

6.2.1 Generative Adversarial Networks (GANs)

GANs are a kind of generative model intended to generate realistic samples of data. They are particularly renowned for producing images and other visual and auditory products that closely resemble real-world data. Figure 6.1 illustrates the GAN training process, where the generator creates images and the discriminator evaluates them, improving over time.

6.2.2 Principles Behind GANs

- **Adversarial Training**:

    - **Generator (G)**: The generator generates synthetic data samples from noise vector z. It is designed to produce data that is as realistic as possible. For example, if z is random noise, $G(z)$ is an image that the generator wants to appear like a real image.
    - **Discriminator (D)**: The discriminator is in charge of differentiating input data that is genuine from the input data that has been produced by the generator. It provides an estimate of a true probability distribution $D(x)$ quantifying the likelihood that a sample x is genuine, a sample originating randomly from the population, not created with G.

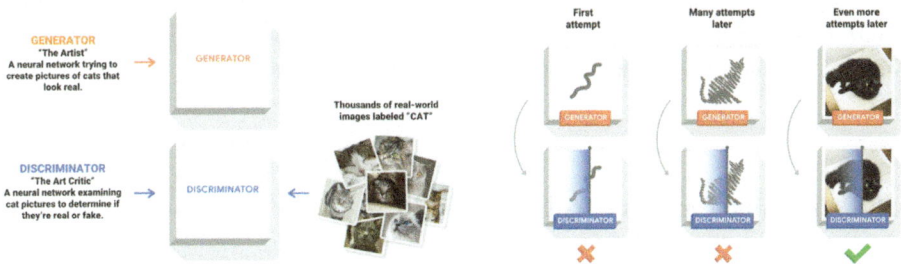

Fig. 6.1 Illustration of the GAN training process, where the generator creates images and the discriminator evaluates them, improving over time [10]

- **Adversarial Game**:

    - **Objective Function**: The generator and the discriminator are trained in opposite manners. The generator's target is to transform the discriminator and make it distinguish fake samples as real ones. On the other hand, the discriminator aims to traverse a real and fake sample class in the best way possible. It is possible to formalize the adversarial setup by the following objective function:

$$\mathcal{L} = \mathbb{E}_{x \sim p_{\text{data}}(x)}[\log D(x)] + \mathbb{E}_{z \sim p_z(z)}[\log(1 - D(G(z)))].$$

where,

 - $p_{\text{data}}(x)$ represents the distribution of real data.
 - $p_z(z)$ is the distribution of the noise input to the generator.
 - $\mathbb{E}$ stands for expectation value regarding the respective distributions.

6.2.3 Structure of GANs

The architecture of GANs consists of two neural networks that are trained simultaneously:

- **Generator Network**: This network generates data samples from random noise, and therefore it can be represented as a function of the noise. Fully connected layers, convolutional layers as well as activation functions make up some of the layers it is generally made of. Ideally, the output of generator should be as close as possible to the distribution of the real data.
- **Discriminator Network**: The discriminator network to decide whether incoming data is real or it is generated. It is often arranged in a manner that consists of the convolutional operations' layers, pooling layers, and fully connected layers. Its output is a probability score of the accuracy of the data that is provided to its input section.

6.2.4 Training Process of GANs

The training process of GANs involves a minimax game between the generator and the discriminator:

6.2 Gain Insight Into Generative Adversarial Networks (GANs)

Fig. 6.2 Adapted from "Unsupervised representation learning with deep convolutional generative adversarial networks [6]"

- **Objective Function**: There is an idea in the generator to make it as difficult as possible for the discriminator to distinguish between the actual data and generated data. At the same time, the discriminator's goal is to make as accurate a distinction as possible. From this flows the following objective function:

$$\mathcal{L} = \min_G \max_D \mathbb{E}_{x \sim p_{\text{data}}(x)}[\log D(x)] + \mathbb{E}_{z \sim p_z(z)}[\log(1 - D(G(z)))],$$

where,

- **pdata(x)** represents the distribution of real data.
- **pz(z)** represents the distribution of the noise input to the generator.
- **E** stands for the expectation value with reference to the respective distributions.
- **G(z)** represents the data generated by the generator from the noise vector z.
- **D(x)** represents the probability that x is real, i.e., a true sample from the data distribution.

- **Training Dynamics**: In their training, the generator and discriminator are trained consecutively as opposed to being trained parallel to each other. The discriminator, which is trained to classify between true and fake data, is trained in a way that increases its accuracy, and the generator, on the other hand, is trained to generate fake data on which the discriminator has a hard time distinguishing from actual data (Fig. 6.2).

6.2.5 Applications in Generating High-Quality Images

It has been observed that in the generation of raw image and other related data, GANs offer a great performance. Their applications include:

- **Image Synthesis**: GANs are able to generate photorealistic images out from noise or from conditions provided to it. They are applied in operations, for instance, image-to-image translation where one type of picture is converted into another, for instance, transforming the sketch into the photo.
- **Example GAN**:

    - **DCGAN (Deep Convolutional GAN)**: It is recommended to start the development of the model at this point and determine its architecture that will allow it to learn and generate realistic images from random noise; one of the most famous types of GANs used at this stage is DCGAN. Convolutional layers are used to extract the spatial information of images. DCGANs have therefore been applied in synthesis of face images, images of animals, and other objects.
    - **Pix2Pix**: Pix2Pix is a type of cGAN especially applied to solve image-to-image translating problems. It can change one type of image to another; for instance, sketch to photo-realistic, black and white pictures to colored, or day shots to night shots.

- **Style Transfer**: GANs are used in style transfer programs, which allow the style of one image to be transferred onto another image while maintaining the main content.

Example GAN:

- **CycleGAN**: CycleGAN is a GAN architecture that is used for style transfer problems, especially for unpaired image-to-image translation. Yes, it can learn one image to mimic the style of another image and in the process, it does not require matched pairs. For example, one aspect will transform a picture of a horse into a zebra or a Monet painting into a photograph.
- **ArtGAN**: ArtGAN is another GAN used in artistic style transfer similar to the context of this paper. It enables the sharing of art styles, for instance, converting a photo to that of a painting and depicting it like that of the famous artists like Van Gogh or Picasso.
- **Data Augmentation**: GANs are employed to create more training data so that the learning models may be enhanced, especially where data are scarce.
- **Example GAN**:

    - **AC-GAN (Auxiliary Classifier GAN)**: To this end, the same AC-GAN is employed to synthesize labeled or augmented data especially when dealing with small datasets. This is a variant of GAN that was designed for generating images that are conditioned on the class labels making it ideal for generating several samples within a given class.
    - **StyleGAN**: StyleGAN is a high-quality GAN architecture that was released by NVIDIA; this architecture is effective in producing highly realistic images while offering a good degree of control over the image style and content. It has been used

6.2 Gain Insight Into Generative Adversarial Networks (GANs)

to generate big databases of synthetic faces, which can be used to extend training sets for such applications as face identification.

- **Super-Resolution**: GANs are useful in increasing the resolution of images by providing high-resolution images from low-resolution images.
- **Example GAN**:

  - **SRGAN (Super-Resolution GAN)**: Specifically, SRGAN is developed to super-resolute the images, meaning the application transforms the images with low resolution to images with high resolution. This is normally used in scenarios that require a high level of image definition, for example, in medical applications and satellite imagery among others.
  - **ESRGAN (Enhanced Super-Resolution GAN)**: SRGAN is an enhanced version of the previous model ESRGAN which is much better than SRGAN that provides more realistic and sharp high-definition images. It includes enhancement such as a better generator and a new loss function for general stability in training and at the same time generating better output (Figs. 6.3, 6.4, 6.5 and 6.6).

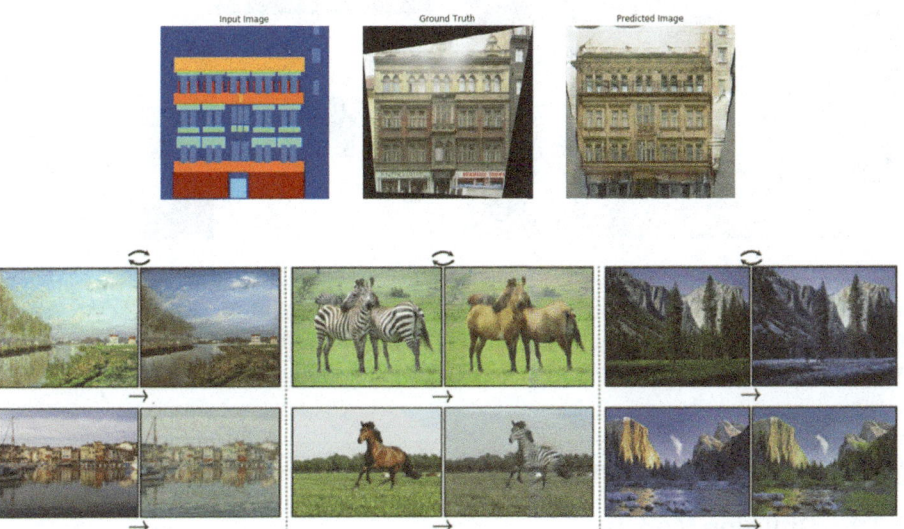

Fig. 6.3 CycleGAN is a GAN architecture used for unpaired image-to-image translation, enabling style transfer between images without requiring matched pairs, such as transforming a horse into a zebra or a Monet painting into a photograph [7]

Fig. 6.4 An example from ArtGAN [9]

Fig. 6.5 Image samples from ImageNet dataset. Samples provided in Salimans et al. [11] which has been produced from the model. Random samples that are being drawn from an AC-GAN are also shown below—Source Random samples from an AC-GAN. While the samples of AC-GAN samples have a global coherence that is not present in the other samples of the earlier model

6.3 Explore Diffusion Models and Their Types

6.3.1 Diffusion Models

Diffusion models are one of the most recent classes of generative models that generate photorealistic samples of data by emulating the process of adding and then removing noise [5]. These models are based on some generic physical processes such as diffusion of ink in water and the principles of information theory. Next, it's time to discuss how diffusion

6.3 Explore Diffusion Models and Their Types

Fig. 6.6 StyleGAN, developed by NVIDIA, generates highly realistic images with control over style and content, aiding in the creation of synthetic face databases for face identification [12]

models operate, the way that they are built, and the steps needed for training them (Figs. 6.7, 6.8 and 6.9).

- **Noise Schedule**: Introducing Noise Gradually
  Concept: You can see, for instance, how one begins with an image, just as though one started with a photograph. So, let's think about including noise in the signal gradually, like in the TV static till you can no longer even identify the shape of the figure showcased. This progressive addition of noise is regulated by what is normally referred to as the "noise schedule."

  - **Noise Levels**: The noise schedule is, in fact, a flow chart, in which noise is introduced into the data at each appropriate point. First of all, the image is visible and then with further steps, more and more noise is added to the picture making it almost nonexistent in the end.
  - **Mathematical Representation**: Thus, this process can be represented with the help of the equation:
    $$x_t = x_{t-1} + \epsilon_t,$$
    where x_t represents the image at time step t and ϵ_t represents the noise added at each step.

Fig. 6.7 SRGAN (Super-Resolution GAN) transforms low-resolution images into high-resolution counterparts, enhancing image definition for applications such as medical imaging and satellite imagery [13]

- **Markov Chain**: The Stepwise Degradation
  Concept: Noise addition can be described as a set of steps starting with one step and until the last one depending on each other. This sequence is known as a Markov chain. In other words, the next condition of the chain depends only on the present condition of the chain and not on the conditions in the past.

  - **Stepwise Process**: At each step a small amount of noise is instilled into the data in a way that degrades it even further. This process goes on until most of the information in the data is noise and non-sensical.
  - **Mathematical Representation**: It is essential to note that each step of the Markov chain will be described:

  $$x_t = x_{t-1} + \epsilon_{t-1}.$$

6.3 Explore Diffusion Models and Their Types

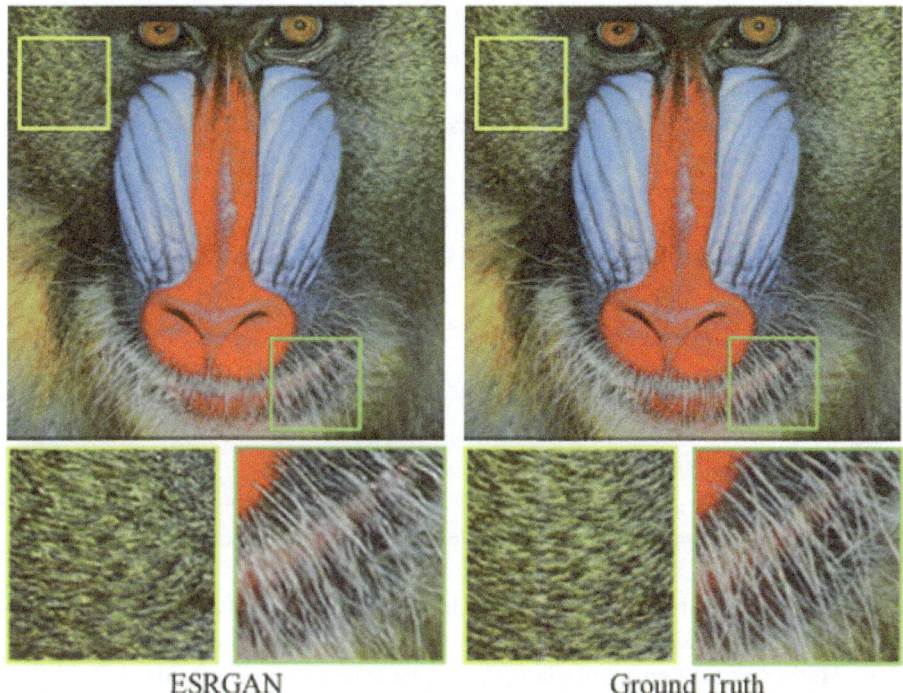

Fig. 6.8 ESRGAN (Enhanced super-resolution GAN) improves upon SRGAN by providing sharper, high-definition images through a better generator and a new loss function for enhanced training stability and output quality [14]

Fig. 6.9 Ink diffusion in water [15]

- **Conditional Modeling**: Learning How Data Transforms
  Concept: Thus, in order for the diffusion models to learn realistic data the models require to understanding of how the data changes as noise is added. This is achieved through conditional modeling in which the model is able to tell what the data were in every stage where noise was introduced.

    - **Conditional Distribution**: The model learns to estimate the appearance of the data at each noise level based on its appearance at the previous level. This is represented as
    $$p(x_t \mid x_{t-1}).$$

  This equation shows that the data at step t is conditioned on the data from the previous step $t - 1$.
- **Reverse Process**: Cleaning Up the Noise
  Concept: Once the data is completely noisy, the model's task is to reverse the process and remove the noise step by step, ultimately reconstructing the original data.

    - **Noise Removal**: Think of this like cleaning up a blurry image. Starting with the noisy image, the model gradually removes the noise to restore the original clarity.
    - **Mathematical Representation**: The reverse process is represented by
    $$x_{t-1} = x_t - \epsilon_{t-1}.$$

  The model works backward, subtracting the noise added at each step to recover the original data.
- **Training Objective**: Optimizing the Model
  Concept: To make the model effective at generating data, we need to train it. The goal is to adjust the model's parameters so that it accurately predicts and reverses the noise addition process.

    - **Likelihood Maximization**: The model is trained to maximize the likelihood of correctly predicting the data at each step. This involves minimizing the difference between the model's predictions and the actual data.
    - **Objective Function**: The training objective can be expressed as
    $$L = \sum_t \mathrm{KL}\left(p(x_t \mid x_{t-1}) \parallel p_{\mathrm{model}}(x_t \mid x_{t-1})\right).$$

  This loss function L sums overall noise levels t to ensure the model learns to accurately predict the data at each stage.

- **Practical Example**: Generating an Image Imagine you start with a random noise image and want to generate a realistic photograph. The diffusion model will

  - **Noise Introduction**: Start with a clear image and add noise step by step until the image is just noise.
  - **Learning Process**: During training, the model learns how the image changes as noise is added.
  - **Noise Removal**: To generate a new image, the model starts with random noise and reverses the process, removing noise to create a clear, realistic image.

6.3.2 Applications of Diffusion Models

Diffusion models have proven useful in several advanced applications:

- Image Generation: Producing realistic images from diverse inputs.
- Text-to-Image Synthesis: Creating images based on textual descriptions, enabling complex visual storytelling.
- Style Transfer: Transferring the artistic style of one image onto another, producing visually compelling results.
- Super-Resolution: Enhancing low-resolution images to higher resolutions, improving clarity and detail.

6.3.3 Architecture of Diffusion Models

The architecture of diffusion models typically involves three key components:

- Latent Representation Model: This neural network encodes an image into a latent representation. The goal is to learn a mapping from images to latent vectors such that similar images have similar representations. The model is trained using maximum likelihood estimation to maximize the probability of real images in the dataset.

$$p(x_0 \mid z) = \mathcal{N}\left(x_0; \mu(z), \sigma^2(z)\right).$$

- Diffusion Process: A Markov chain that adds noise to the latent representation, gradually increasing the noise level. This process is often modeled using a Gaussian diffusion approach:

$$p(x_t \mid x_{t-1}) = \mathcal{N}\left(x_t; \mu_t(x_{t-1}), \sigma_t^2\right).$$

The diffusion rate parameter controls the amount of noise added.

- Decoding Process: This neural network reconstructs the image from the latent representation. It is typically trained using mean squared error (MSE) loss to minimize the difference between the generated and original images.

$$L_{\text{MSE}} = \frac{1}{N} \sum_{i=1}^{N} (x_i - \hat{x}_i)^2.$$

6.4 Types of Diffusion Model

LEANs are one broad class of generative models, of which there are many different forms, each of which functions differently in data generation. These models include the idea of how these architectures and specific implementations have been applied in the development of deep learning to give real images. We now look at the general overview of how they work and how they are performed by active agents.

6.4.1 Denoising Diffusion Probabilistic Models (DDPMs)

Deep denoising diffusion probabilistic models or DDPMs are currently employed in image synthesis with synthesis quality. Especially, they are intended to produce realistic images by learning how to invert a process which progressively adds noise to data (Figs. 6.10 and 6.11).

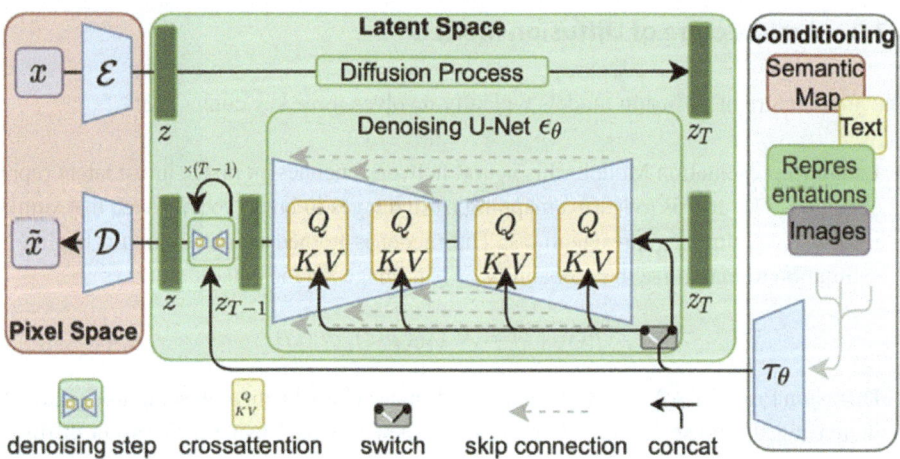

Fig. 6.10 Illustrates latent representation model in diffusion models [16]

6.4 Types of Diffusion Model

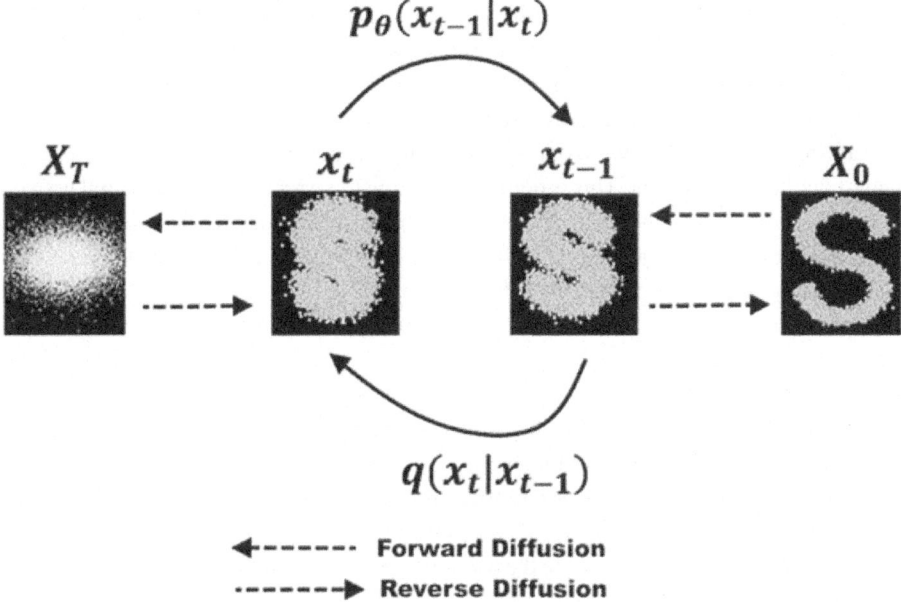

Fig. 6.11 Diffusion process in diffusion models [8]

- **Core Concept**
  The main concept that lies in the background of DDPMs is the modeling of the distribution of the data via diffusion. This process includes two primary steps:

  - **Forward Diffusion Process**: This is a Markovian process where more and more noise is introduced to the original data such as an image. De-coupled here are the steps of mapping the original image to a noise distribution and then to another noise distribution. In terms of mathematical representation, forward process can be represented with a sequence of stochastic operations controlled by the noise schedule.
  - **Reverse Diffusion Process**: Based on the "forward process," which is learned by DDPMs, they work to reverse it. Accordingly, we begin with a noisy image and gradually clean it to obtain a data distribution losslessly with several processing rounds. The reverse process is learnt from a neural network that maps the noise at each level in order to reconstruct images from just noise.

- **Mathematical Framework**

  In the forward diffusion process, data is diffused as per a distribution $q(x_t \mid x_{t-1})$, where x_t is the noise image at step t. It includes adding Gaussian noise, and the objective is to describe this forward diffusion process in T steps.

The reverse process is represented by $p_\theta(x_{t-1} \mid x_t)$, wherein θ refers to the parameters of the neural network. The model can predict which amount of noise is added at each level, and then it starts the process of denoising this image. The last step is to reconstruct the actual image from noise, as it is done using the reverse process T times.
- **Training** Training DDPMs requires to minimize a variational bound on the negative log-likelihood of the data. This is done by adjusting the parameters of the neural network to study and estimate the noise added during the forward process. This used to be done through terms in the objective function which compel the model to generate images that are as close as possible to the true data distribution.
- **Applications** DDPMs have been used in several applications, which show a good performance of the models:

  - Image Synthesis: Generating high-resolution images that are visually appealing and closely resemble real-world images.
  - Inpainting: Filling in missing parts of an image in a coherent and contextually relevant manner.
  - Super-Resolution: Enhancing the resolution of low-resolution images while preserving details.

- **Advantages and Challenges**

  - Advantages:

    - High Quality: DDPMs often produce high-quality images with fine details, surpassing many other generative models.
    - Flexibility: They can be adapted for various generative tasks beyond image synthesis.

  - Challenges:

    - Computational Cost: The iterative nature of the reverse diffusion process can be computationally expensive, requiring many steps to achieve high-quality results.
    - Training Complexity: Training DDPMs involves complex optimization and large datasets, making it resource intensive.

6.4.2 Score-Based Diffusion Models (SBMs)

Score-Based Diffusion Models (SBMs) are a type of generative model used to produce high-quality data samples by learning the underlying distribution of a given dataset. They are part of a broader category of models known as diffusion models, which generate data through a series of gradual transformations from noise to data.

6.4 Types of Diffusion Model

- **Core Concept**
  Score-based diffusion models operate by learning a score function that guides the diffusion process. This score function, often referred to as the score function or gradient of the log density, is used to estimate the gradient of the data distribution's log-likelihood. The core idea is to iteratively refine noisy samples to approximate data samples that resemble the training distribution.
- **Key Components**

  - **Diffusion Process**: The process starts by adding noise to the data through a series of steps. This transforms the data into pure noise. The objective to approach and overcome this process, by transforming noise back into data. Forward diffusion process basically entails the process of sequentially degrading the data with noise for a number of iterations.
  - **Score Function**: To do that, the score function is used during the reverse process in order to estimate the gradient of the data distribution. Despite the fact that the noisy samples often contain a lot of noise, this function aids in reversing the diffusion process so that the samples are directed back to the data distribution.
  - **Training Objective**: The training aim is to find the values of both input parameters that will reduce the squared differences between the predicted score function and the true score function to the smallest possible value. This is done in most cases by minimizing an optimization function, which quantifies the difference between the estimated and true gradients of the data density.
  - **Sampling**: After training, it is possible for an SBM to sample new data, by beginning with noise and following the reverse of the learned diffusion processes. The score function improves the noisy samples and gives high-quality data while in the iteration process.

- **Advantages**

  - **High-Quality Samples**: SBMs can produce high-fidelity samples due to their effective learning of the data distribution.
  - **Flexibility**: They can model complex distributions and are not limited by specific assumptions about the data structure.

- **Challenges**

  - **Computational Complexity**: The iterative nature of the diffusion process can be computationally intensive.
  - **Training Difficulty**: Training score functions can be challenging, especially in ensuring that the estimated scores are accurate and converge properly.

- **Applications**

  - Image Generation: Studies have revealed that the use of SBMs is effective in learning complex visual distributions in the generation of high-quality images.
  - Data Augmentation: They can be used to create synthetic data that can be applied to train other models making the training data more diverse.

6.5 The Logistics of Understanding DALL-E 2

DALL-E 2 is a new kind of model by OpenAI that produces photorealistic images to textual descriptions. A brief knowledge of this technology precisely entails understanding CLIP training and the text-to-image generation process. Here's a detailed explanation of how DALL-E 2 functions.

- **Input Text Processing**
  DALL-E 2 starts with textual supervision where users provide captions that describe the images which are wanted. These descriptions are useful as they serve as a framework for creating content graphical in nature as a result.
- **Encoding Using CLIP**
  Moreover, the input text is taken through the CLIP neural networks, a model that can compare text and images hence translating text into vectors. This process converts textual descriptions into CLIP text embeddings:

$$\text{CLIP\_text\_embedding} = f_{\text{CLIP}}(\text{text}),$$

where f_{CLIP} is a function for encoding the contents of the CLIP model.
CLIP Training: CLIP learns how to align images and texts to one common space that allows the comparison and alignment of the two modalities.

$$\text{Similarity} = \cos(\text{CLIP\_text\_embedding}, \text{CLIP\_image\_embedding}),$$

where cosine similarity determines the distance of the two text and image embeddings in the common space.
- **Conversion to CLIP Image Embeddings through Prior**
  Subsequently, a "Prior" model, which may be an autoregressive model or a diffusion model, is applied to the CLIP text embeddings. The Prior is responsible for converting the textual embeddings into CLIP image embeddings:

$$\text{CLIP\_image\_embedding} = g_{\text{Prior}}(\text{CLIP\_text\_embedding}),$$

6.5 The Logistics of Understanding DALL-E 2

where g_{Prior} is the prior model used according to the following specifications.

- **Diffusion Model**: Concretely, in the case of DALL-E 2, this conversion is performed by using the diffusion model as it has proven useful in the synthesis of high-quality images. The diffusion model reduces noise systematically from images, and in each step, it refines image representations progressively.

- **Final Image Generation**
Afterward, the CLIP image embeddings go to the diffusion decoder for diffusion map generation. This decoder translates these embeddings into the final image:

$$\text{Final\_image} = h_{\text{decoder}}(\text{CLIP\_image\_embedding}),$$

where h_{decoder} depicts a decoding function. Based on some probability distribution, the h_{decoder} can be defined.

- **Decoding Process**: The diffusion decoder takes in the CLIP image embeddings and outputs an image in the goal of creating an image that matches the description given.
- **Effectiveness of Integration**: When designing and training DALL-E 2, it was discovered that using a Prior—like a diffusion model before decoding—gave improved results as opposed to utilizing CLIP text embeddings in the decoder directly.

According to the visual diagram given below (Fig. 6.12), the following concepts are explained:

- **Top Part: CLIP Training Process**
The upper portion of the diagram illustrates the CLIP training process, which involves:

  - Shared Representation Space: CLIP learns a joint representation space for both text and images. This space allows for meaningful comparison and relationship between textual and visual data.

- **Bottom Part: Text-to-Image Generation Process**
The lower part of the diagram demonstrates the process of converting text to images using DALL-E 2:

  - Text Input Encoding: The textual description is encoded into CLIP text embeddings using the CLIP encoder.
  - Prior Processing: These embeddings are processed through the Prior model to generate CLIP image embeddings.
  - Image Decoding: The CLIP image embeddings are then decoded by the diffusion decoder to produce the final image (Fig. 6.12).

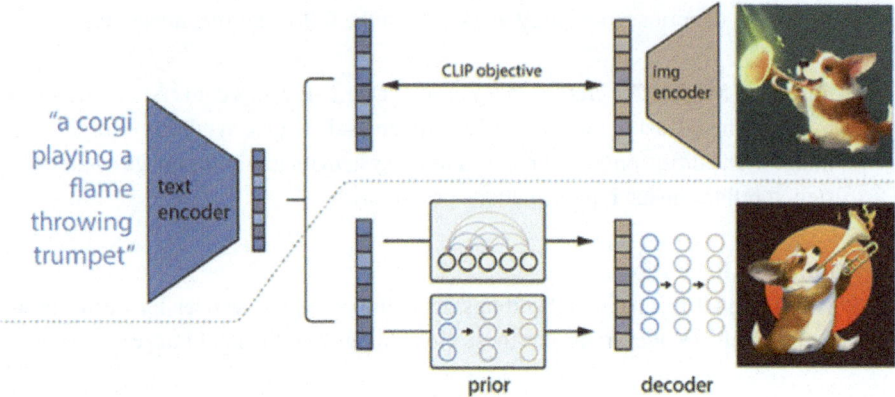

Fig. 6.12 Illustrates DALL-E 2 [17]

6.6 Learn About Stable Diffusion and the Latent Diffusion Model (LDM)

Stable diffusion leverages the Latent Diffusion Model (LDM) to efficiently generate high-quality images. Understanding this technology involves exploring several core concepts and components.

6.6.1 Latent Diffusion Model (LDM)

Based on the Stable Diffusion framework, LDM stands for central to its integration into the methodology. It extends diffusion models, which operate in the latent space of autoencoders. Here's a breakdown of the technology:

- **Diffusion Models in Latent Space**

  - **Diffusion Process**: Conventional diffusion models, on the other hand, successively add noise to data and afterward try to recoup the data. LDM has adopted the same approach but to work in the latent space instead of directly working on the input data, such as image data. The process of diffusion has been further revised to include the addition of noise in the latent representations of the data.
  - **Latent Space Application**: In LDM, the noise is added to the compressed latent representations instead of the raw input data, which means

    $$\text{Latent\_noisy} = \text{Latent} + \epsilon,$$

where ϵ represents the noise added during the diffusion process.

- **Autoencoders and Latent Representations**

  - **Autoencoders**: These are neural networks that are used in the encoding of data at the input layer and the generation of the encoded feature space and decoding of this latter feature space into the original output space. In LDM, pre-trained autoencoders generate a latent productive space that encodes a number of necessary attributes of the input data.
  - **Latent Space Utilization**: The latent space provides a compact representation of the data, enabling more efficient processing:

  $$\text{Latent\_representation} = f_{\text{encoder}}(\text{Input\_data})$$

  $$\text{Reconstructed\_data} = f_{\text{decoder}}(\text{Latent\_representation}),$$

  where f_{encoder} and f_{decoder} are the encoding and decoding functions of the autoencoder, respectively.

- **Training and Optimization**

  - **Training Objective**: LDMs are trained to model the diffusion process within the latent space. The objective is to optimize the model parameters so that the transformation from noisy to clean latent representations is accurately learned.
  - **Optimization Goal**: The training process minimizes the difference between the noisy and clean latent representations, adjusting the model to effectively handle noise introduction and denoising.

- **Cross-Attention Layer**

  - **Incorporation**: LDMs use a cross-attention layer to enhance their ability to manage various conditional inputs, such as text descriptions and bounding boxes.
  - **Function**: This layer improves the model's capacity to generate high-resolution images by allowing it to focus on different parts of the input data, facilitating

  $$\text{Attention\_output} = \text{softmax}(W_{\text{cross}} \cdot \text{Latent\_representation} + b_{\text{cross}}),$$

  where W_{cross} and b_{cross} are the weight and bias parameters of the cross-attention mechanism (Fig. 6.13).

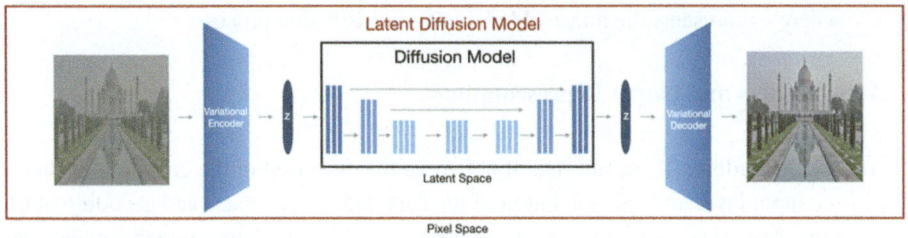

Fig. 6.13 Latent diffusion model [18]

6.6.2 Benefits and Significance

- **Computational Efficiency**: By operating in the latent space of pre-trained autoencoders, LDMs reduce the computational resources needed for training diffusion models, making them more resource efficient.
- **Complexity and Fidelity**: Training diffusion models within the latent space allows LDMs to balance between simplifying representations and preserving detailed information, which enhances the visual fidelity of the generated images.
- **Conditioned Synthesis**: The cross-attention layer allows LDMs to conditionally generate images based on various inputs, such as text, increasing their versatility in image synthesis.

Stable diffusion is based on the Latent Diffusion Model (LDM) that provides effective and high-quality image synthesis. Therefore, the use of diffusion models in autoencoders' latent space and methods like cross-attention makes LDMs one of the latest advancements in AI image synthesis. This approach saves computational time, preserves image quality, and facilitates the imparted conditioned synthesis, which is a great advancement in generative AI systems.

6.7 Find Out What Awaits You During Midjourney: Learn the Technology Behind Midjourney

Midjourney employs complex generative procedures and approaches when translating the textual cues into aesthetically appealing images. Midjourney—The following represents an evaluation of the current technological platform employed by the company:

6.7.1 Generative Adversarial Networks (GANs)

- Architecture: GANs consist of two primary components:

  - Generator: Creates images from random noise.
  - Discriminator: Differentiates between real images and those produced by the generator.

- Adversarial Training: This setup leads to an adversarial process where:

  - Generator's Objective: Enhance its quality of producing images that can easily fool the discriminator and think that the images are real data samples.
  - Discriminator's Objective: Improve its capacity of discriminating generated and true images.

- Such competitive nature pushes both the networks to improve their performances with each iteration resulting into more realistic images.

6.7.2 Text-to-Image Synthesis with GANs

- Conditioned Generation: Midjourney uses GAN architecture in the creation of image from text descriptions.
- Text Conditioning: Text inputs are conditioned to the generator so as to generate appropriate images that correspond with the descriptions given.
- Latent Representation: The text is represented using a lay hidden dense vector that controls the generation process of images.

$$\text{Latent\_vector} = f_{\text{encoder}}(\text{Text\_description}).$$

- Image Generation: It aids the generator in creating images of the right nature in order to match the contents of the text prompt.

6.7.3 Conditional GANs (cGANs)

- Enhanced Conditioning: Midjourney, however, utilizes a derivative called Conditional GANs (cGANs).

- Conditional Inputs: They both take extra input in the form of information on the generator side and the discriminator side such as the text description. This additional conditioning enhances the correlation that exists between the generated image and the text input.

$$\text{Generator\_output} = G(\text{Latent\_vector}, \text{Text\_description})$$

$$\text{Discriminator\_score} = D(\text{Image}, \text{Text\_description}).$$

- Improved Alignment: This approach improves the generator's capacity to generate images that are relevant to given text descriptions.

6.7.4 Training Process

- **Iterative Updates**:
  The training includes the following:

  - **Generator**: It seeks to generate images that are as close to real as possible in order to fool the discriminator.
  - **Discriminator**: Attempting to classify real images from fake images. As machine learning models are constructed to intelligently distinguish between real and generated images.

- **Loss Functions**:

  Both components are optimized using the L1 or L2 or any other loss functions that point out the areas which need to be improved.

$$\text{Loss}_{\text{gen}} = -\log(D(G(z, \text{text})))$$

$$\text{Loss}_{\text{disc}} = -\log(D(\text{real})) - \log(1 - D(G(z, \text{text}))),$$

where D stands for discriminator, G for generator, and z as random noise.

6.7.5 Attention Mechanisms

- **Contextual Focus**: The attention mechanisms may be used to do the following.

  - **Enhance Focus**: Optimizing the areas of the image to which the generator pays more attention depending on the input text.

- **Selective Emphasis**: Let the model be selective on which part of the image to generate, to generate more semantically meaningful images.

$$\text{Attention\_weights} = \text{softmax}(W \cdot \text{Latent\_representation} + b),$$

where W and b are the parameters of the attention mechanism.

6.7.6 Data Augmentation and Preprocessing

- **Data Augmentation**: Any methods that will enhance the size of the training dataset and reduce on the generality of the model may be utilized.
- **Text Preprocessing**: Contents of textual descriptions are preprocessed in certain ways such as tokenization, they help to transform the text in the appropriate form for the model:

$$\text{Text\_embedding} = \text{Embedding\_function}(\text{Tokenized\_text}).$$

6.7.7 Benefits and Applications

The technology behind Midjourney enables the creation of realistic and contextually relevant images based on textual descriptions:

- **Realistic Image Synthesis**: The ability to produce imagery from text generates Midjourney as being very functional in many sectors, such as graphic design, media writing, and imagination.
- **Innovative Applications**: The integration of language and image synthesis in Midjourney creates possibilities for new approaches in generative AI's application based on the domain, and the results are unmatched to anything that can be made conventionally in the digital art world.

6.7.8 Example

- Input Text: A lovely picture which depicts a calm sunset over the mountains.
- Text Encoding: The text is passed through a process that produces a latent representation that represents a futuristic city.

Fig. 6.14 A serene sunset over a mountain range (created by the authors)

- Image Generation: In the case of Midjourney, the result of cGANs is an image for which neon lights and skyscrapers are emphasized in the picture which, according to the text description, should be.
- Attention Mechanism: This aspect also points at specific parts or attributes such as the lighting effects to make sure they depict what has been described.
- Output: The last picture (Fig. 6.15) illustrates a night city with lighted tall buildings, which will also underline input text as futuristic.

Figure 6.14 is a depiction of an image based on a description: "A serene sunset over a mountain range." This image depicts the beauty of the sunset with warm colors and a mountain range in the background.

Midjourney uses Generative Adversarial Networks (GANs), more so Conditional GANs (cGANs), to generate images from text descriptions. Namely, by training both the generator and the discriminator on textual inputs, Midjourney improves the correspondence between generated image and textual description. The training for several iterations, incorporation of attention mechanism, and enhanced data preprocessing make it produce high-resolution and semantics relevant image.

6.8 Why DALL-E 2, Stable Diffusion, and Midjourney Are Not The Same Thing?

A comparison between DALL-E 2, Stable Diffusion, and Midjourney provides important insights into the current position of text-to-image generation models. In the subsequent section, the characteristics of every model and its approaches will be described, and similarities and differences between them will be discussed.

6.8.1 DALL·E 2: An Improved Text-to-Image Translation Using CLIP and Diffusion Models

- **Core Technology**: DALL-E 2 uses CLIP as well as diffusion models which are utilized in elaborating images from the text input.
- **CLIP Integration**: CLIP is used by DALL-E 2 to embed both text and image into vectors in a high-dimensional space to map semantics of textual descriptions to the image space.

$$E_{text} = \text{CLIP}_{text}(T), \quad E_{image} = \text{CLIP}_{image}(I).$$

Here, Etext is the CLIP embeddings for text and Eimage is for the images.

- **Diffusion Process**: It uses a function called diffusion model that iteratively adds and removes some random noise to an image to produce a high-quality image from text input. This process is performed under the consideration of a learned probability distribution which directs the generation.

$$E_\theta(x_{t-1} \mid x_t) = \mathcal{N}(x_{t-1}; \mu_\theta(x_t, t), \Sigma_\theta(x_t, t)).$$

- **Strengths**: Here, Etext is the CLIP embeddings for text and Eimage is for the images.
- **Diffusion Process**: It uses a function called diffusion model that iteratively adds and removes some random noise to an image to produce a high-quality image from text input. This process is performed under the consideration of a learned probability distribution which directs the generation.
- **Generated Image**: DALL-E 2 might create a realistic depiction of the text prompt, for example, neon-lit skyscrapers reflected on wet pavement with a lot of details, to an extent, at which it is difficult to distinguish the image from a photograph.

6.9 Stable Diffusion: Latent Diffusion Models (LDMs) Toward Making the Business More Efficient

- **Core Technology**: Stable diffusion builds upon Latent Diffusion Models (LDMs), which are a form of diffusion models that work in the latent space of well-established autoencoder networks.
- **Latent Space Diffusion**: Stable diffusion uses the diffusion process in a compressed latent space where by this, the transformation of data in the latent space is made efficient by the fact that a lot of computational resources are conserved while maintaining the quality of the images (Fig. 6.15).

$$z = E_{\text{latent}}(x) \quad \text{and} \quad \hat{x} = D_{\text{latent}}(z).$$

Here, z represents the hidden variables and $\hat{x}$ is the generated image.

- **Cross-Attention Mechanism**: The LDMs in Stable Diffusion are designed with cross-attention layers that enable the model to derive more specialized information such as text descriptions and generate detailed images.

$$\text{Attention Output} = \text{softmax}\left(\frac{QK^T}{\sqrt{d_k}}\right)V.$$

Fig. 6.15 Generated image by DALL-E 2 depicting neon-lit skyscrapers reflected on a wet pavement, showcasing intricate details that make it hard to distinguish from a photograph (created by the authors)

Fig. 6.16 Another view of a human and a futuristic robot who are challenged to a game of chess (created by the authors)

- **Strengths**: As it has been pointed out, Stable Diffusion is computationally inexpensive and the model is easily trainable and runable with relatively lower end hardware while achieving quality results. Latent space also enhances the model's capacity to fine-tune the level of detail and level of complexity in the manufactured images.
- **Applications**: That is why it is ideal for applications with limited computational capabilities like the generation of synthetic images in real time, as well as tasks that require precise control over images' characteristics.
- **Generated Image**: An extremely well done, fully realized picture of an elegant, shiny newfangled robot playing chess with some human and depicting their facial emotions and staging (Fig. 6.16).

6.10 Midjourney: Here, Creative Synthesis with Conditional GANs (cGANs)

- **Core Technology**: Midjourney uses cGANs, the extension of the base GAN model that incorporates additional conditionality for both the generator and the discriminator and in the given case, the textual description.
- **Conditional GAN Architecture**: In Midjourney, the generator is conditioned on text which means that it will only produce the image that corresponds to the description provided. The discriminator also determines the legitimacy of these images with focus made on the conditioning text.

$$\text{cGAN Objective}: \quad \min_G \max_D V(D, G \mid c).$$

- **Attention Mechanisms**: Midjourney could use the attention mechanisms to direct the model's computation on generation parts to more appropriately match image context.

$$\text{Attention Output} = \text{softmax}\left(\frac{QK^T}{\sqrt{d_k}}\right).$$

- **Unique Strengths**: Midjourney relies on the adversarial training framework of cGAN and hence develop the model that creates incredible images with unique artistic appeal. The model is outstanding at producing the artistic representation of given texts and often leans toward stylistic representations.
- **Applications**: The Midjourney is most suitable for creative businesses, for example, graphic design, arts, content creation, and other sectors where style over substance is preferred especially when it comes to graphical designs.
- **Generated Image**: An artistic and imaginative interpretation of a fantastical landscape, with floating islands and cascading waterfalls, rendered in a visually striking style (Fig. 6.17 and Table 6.1).

Fig. 6.17 Text prompt: "A place that looks like planets, islands which are floating and has waterfalls" (created by the authors)

Table 6.1 Summary of differences

Aspect	DALL-E 2	Stable diffusion	Midjourney
Technological foundation	Combines CLIP and diffusion models for semantically accurate image generation	Uses latent diffusion models (LDMs) in latent space for computational efficiency	Relies on conditional GANs (cGANs) for creative and artistic outputs
Strengths	Excels in producing detailed and conceptually accurate images	Balances complexity with computational efficiency, delivering high-quality images with fewer resources	Focuses on creativity and stylistic flair, producing visually striking and imaginative images
Applications	Ideal for tasks requiring detailed realism, such as concept art and illustration	Optimized for resource-efficient applications, such as real-time image synthesis	Best suited for creative and artistic endeavors, including graphic design and visual arts

6.11 Understand the Role of Data Augmentation and Preprocessing

Data augmentation and preprocessing are critical steps in enhancing the performance of diffusion models and other generative AI systems for image synthesis. These techniques ensure that models are well trained, robust, and capable of producing high-quality outputs. Below, we delve into the significance of these processes, explaining how they contribute to the overall effectiveness of image generation.

6.11.1 Data Augmentation: Enhancing Model Generalization

Definition: Data augmentation involves artificially expanding the training dataset by applying a variety of transformations to the original data. These transformations can include:

- Rotations: Rotating the image by a certain degree.
- Flips: Mirroring the image horizontally or vertically.
- Translations: Shifting the image along the x- or y-axis.
- Scaling: Resizing the image.
- Color Adjustments: Modifying the brightness, contrast, saturation, or hue.

Purpose: Data augmentation has a different aim; it is to make the training data more diverse, although new data is collected. This is done by feeding the model with nearly similar data but slightly different from the original, and as a result improves the model's ability in generalization.

$$\text{Augmented Data} = \{T_i(x) \mid T_i \in T, x \in \text{Training Set}\}, \quad (6.1)$$

where $T_i(x)$ is a rewrite of an image x by the transformation T_i and T is the set of all the possible transformations.

Impact on Performance: The use of data augmentation also assists in minimizing overfitting where the model is very good in the training data but very poor in new data. If the model learns from other variations in the dataset, it becomes not easily affected by certain changes in the input variable which leads to more real-life performance.

Example in Diffusion Models: In diffusion models, the augmented data may help to improve the model's capacity to create more realistic images through training it with various image conditions. It is especially crucial when speaking of the diffusion process, which implies the gradual addition and reducing of noise and how effective it can work in different cases.

6.11.2 Preprocessing: Cleaning Data for the Best Model Training

The process of data cleaning is a process through which data that has been collected is prepared before it is taken through a model. Some of the preprocessing that can be applied are normalization, scaling, removing noise, and converting textual descriptions to numerical patterns.

Normalization: Normalization of images involves scaling the pixel values so that they lie between 0 and 1 or -1 and 1. This helps stabilize the input data, which enhances the speed of training. This is similar to the z-score in statistics.

$$x_{\text{normalized}} = \frac{x - \mu}{\sigma}, \quad (6.2)$$

where μ represents the mean and σ represents the standard deviation of the pixels of the dataset.

Resizing: In most cases, the accurateness of the images is regulated by making them occupy a similar physical size so they are standardized in terms of the size of a batch convenient for the training.

Denoising: For generative models and more specifically for diffusion-based ones, denoising constitutes a preprocessing step. The process of denoising helps the model understand how clean images should look from the noisy inputs, which is essential when generating new samples in diffusion models:

$$\hat{x} = D_\theta(x + \epsilon), \quad \epsilon \sim \mathcal{N}(0, \sigma^2),$$

where $\hat{x}$ is the denoised image, D_θ is the denoising function that is decided by the parameter θ, and ϵ is the noise added to the original image x.

Text Embeddings: For the generative models that create images from text, there is also the preparation of text into vector representations referred to as embeddings.

$$E_{\text{text}} = \text{Embed}(T),$$

where T represents the text input and E_{text} represents the respective texture embedding.

6.11.3 Synergy Between Augmentation and Preprocessing

Combined Impact: It becomes evident that combination of data augmentation with preprocessing has a positive impact on model performance. Augmentation adds randomness in the training data and hence the model becomes more resilient while preprocessing helps in having data in an appropriate form for learning.

- **Training Stability**: There are certain measures such as normalization or resizing in order to optimize the training process since the model will not be subjected to small data fluctuations during its training process.
- **Quality of Generated Images**: It clearly demonstrates that better preprocessing working in parallel with reliable data augmentation would actually bring benefits to the quality of the generated images. This means that the model is able to manage the given input conditions well enough and thus produces enhanced precision and graphic display results.

Augmentation and preprocessing of data are methods that are very helpful while training, developing, and creating diffusion models and generative AI for image synthesis. These techniques increase the robustness, generalization, and quality of the generated images by absorbing fivefold of initial training data and taking necessary format of input data. In particular, the effectiveness of augmentation with preprocessing emphasizes the influence of both concepts on the development of new AI models that are more efficient and effective in solving problems.

6.12 Explore the Use of Attention Mechanisms in Image Generation

This is because attention mechanisms have transformed how deep learning is done especially in data that is sequential such as text. Similarly, in the case of attention mechanism, it has been equally revolutionary specifically in image generation in which the model is provided with the capability to make coherent images that are in line with the textual descriptions. In this work, we discuss the implications of attention mechanisms to the general image synthesis process and further their ability to augment text-to-image models.

6.12.1 Understanding Attention Mechanisms

- **Definition**: Full-sentence/contextual attention mechanisms permit a model to pay more attention to some parts of input data compared to others giving higher importance to regions/tokens. This dynamic allocation of focus helps the model to increase the nominees of relevant features for the task.
- **Self-Attention**: In self-attention, both the keys and the values come from the same input that is segmented into different positions where the model aims to capture a relationship. This is important in order to determine the associations in data like the role played by a word in a sentence or position of a pixel within an image.

$$\text{Attention Output}(Q, K, V) = \text{softmax}\left(\frac{QK^T}{\sqrt{d_k}}\right) V,$$

where Q, K, and V are the query, key, and value matrices, respectively, and d_k is the dimensionality of key vectors.

Cross-Attention: Cross-attention is a type of attention where one sequence is the source while the other is the target sequence to be generated such as text while generating an image. This proves particularly handy in text-to-image models where the image synthesis process is controlled by a text description:

$$\text{Attention Output}(Q, K_{\text{text}}, V) = \text{softmax}\left(\frac{QK_{\text{text}}^T}{\sqrt{d_k}}\right) V.$$

6.12.2 Enhancing Image Generation with Attention

Guided Image Synthesis: About text-to-image models, SGAMs ensure that several input text features affect specific areas of the generated image. For instance, the text used might be—"a red apple on a green table," attention enables the model to work on sampling a red color for the apple and green background for the table.

- **Focus on Relevant Features**: The attention mechanisms assist the model on which features to focus on when creating something. This makes sure that the given image can convey the meaning or the semantics of the input text desired, thus gathers more coherent and semantically related results.
- **Spatial Coherence**: As attention mechanisms keep attention tokens applied to some region of the image and the tokens applied to the other regions of the image separate while correlating them to ensure spatial consistencies in the generated image, such mechanisms are beneficial since they reduce the dependence of the transformer on location tokens.

This is important in realistically modeling scenes where objects do not distort and retain their physical proportions and location to other objects and features in the scene.

$$I_{\text{generated}} = \sum_{i=1}^{n} \alpha_i \cdot V.$$

6.12.3 Attention Mechanisms in Diffusion Models

Integration in Diffusion Models: In diffusion models, attention mechanisms are integrated to enhance the denoising process by selectively focusing on relevant parts of the image. This makes it possible for the model to have enhanced image generation whereby the image is improved at each stage of diffusion as seen in the diffusion map.

Contextual Refinement: Attention mechanisms enable diffusion models to refine images contextually, focusing on correcting specific regions that might have been distorted during the noise introduction phase.

$$x_{t-1} = \text{Attention}(\epsilon_\theta(x_t, t, c)) + \sigma_t \cdot \epsilon.$$

6.12.4 Benefits of Attention Mechanisms in Text-to-Image Models

- **Improved Text-Image Alignment**: Attention mechanisms significantly enhance the correlation between the text and the output image. This is important in the creation of images for reflecting the understanding of textual descriptions that may contain complicated details.
- **Versatility in Generation**: Through attention mechanisms, it becomes possible to cover various and rather intricate text prompts and generate images that not only appear realistic but are also rather artistic in terms of the input text descriptions.
- **Reduced Artifacts**: By directing attention to the right areas of the image, attention mechanisms assist in the reduction of artifacts and inconsistencies hence improving on the quality of the images.

6.13 Learn the Key Evaluation Metrics and Optimization Methods

For getting high-quality results in generative models like GAN and others, it is essential to understand about the loss functions and the optimization. These components define how the training happens without the models producing unrealistic and realistic outputs. Here, the choice of the loss functions and optimization techniques used in these models and how they affect the effectiveness of generative models will be discussed.

6.13.1 Role of Loss Functions in Generative Models

Role of Loss Functions: Several types of loss functions act as a significant aid to help generative models identify the gap between the generated output and the expected results. They interact to supply feedback that influences on the model about how best it should change some of its parameters to enhance and gain better performance.

- **Minimization Objective**: In the context of GANs, the generator tries to minimize a certain loss function to produce outputs that are more realistic to real data while the discriminator, on the other hand, aims at maximizing the same function so that he may better recognize between fake and real data. This form of relationship puts pressure on both components to evolve in an incremental manner.

  In the context of GANs, the generator tries to minimize a certain loss function to produce outputs that are more realistic to real data while the discriminator, on the other hand, aims at maximizing the same function so that it may better recognize the difference between fake and real data. This form of relationship puts pressure on both components to evolve in an incremental manner:

  $$\text{Minimize } \mathcal{L}_G = -\mathbb{E}_{x \sim \text{real data}}[\log D(x)] + \mathbb{E}_{z \sim \text{noise}}[\log(1 - D(G(z)))],$$

  where $D(x)$ is the discriminator's prediction for real data x and $G(z)$ represents the generator's output from input noise z.

- **Variational Autoencoder (VAE) Loss**:
  For VAEs, the loss function combines a reconstruction loss with a regularization term to ensure that the latent space is continuous and facilitates smooth interpolation between data points:

  $$\mathcal{L}_{VAE} = \mathbb{E}_{x \sim \text{data}}[\log p(x|z)] - \text{KL}(q(z|x) \| p(z)),$$

  where KL represents the Kullback–Leibler divergence, which regularizes the latent space distribution.

6.13.2 GAN-Specific Loss Functions

- **Binary Cross-Entropy Loss**: The standard loss function used in GANs is the binary cross-entropy loss, where the discriminator is trained to maximize the probability of correctly classifying real and generated samples.

The loss can be expressed as

$$\mathcal{L}_D = -\mathbb{E}_{x\sim\text{real data}}[\log D(x)] - \mathbb{E}_{z\sim\text{noise}}[\log(1 - D(G(z)))],$$

where $D(x)$ is the discriminator's prediction for real data x and $G(z)$ is the generator's output from input noise z.
- **Generator's Loss**:

The generator's loss is designed to fool the discriminator, so it minimizes the likelihood that the discriminator correctly identifies generated data as fake:

$$\mathcal{L}_G = -\mathbb{E}_{z\sim\text{noise}}[\log D(G(z))].$$

- **Discriminator's Loss**: Conversely, the discriminator's loss encourages it to correctly distinguish between real and generated samples. This is often expressed as

$$\mathcal{L}_D = -\mathbb{E}_{x\sim\text{real data}}[\log D(x)] - \mathbb{E}_{z\sim\text{noise}}[\log(1 - D(G(z)))].$$

- **Wasserstein Loss**:

The Wasserstein GAN (WGAN) introduces a different loss function, which aims to improve training stability by using the Earth Mover's Distance (also known as Wasserstein distance). This distance measures the amount of work required to transform one distribution into another.

$$\mathcal{L}_{WGAN} = \mathbb{E}_{x\sim\text{real data}}[D(x)] - \mathbb{E}_{z\sim\text{noise}}[D(G(z))].$$

In WGANs, the discriminator (often called the critic) is not trained to classify inputs as real or fake but to output a value that approximates the Wasserstein distance:

6.13.3 Optimization Techniques

- **Stochastic Gradient Descent (SGD)**: One of the foundational optimization algorithms, SGD updates the model parameters by computing gradients of the loss function concerning the parameters and moving in the opposite direction to minimize the loss.

The update rule for SGD can be expressed as

$$\theta_{t+1} = \theta_t - \eta \nabla_\theta L(\theta_t),$$

where θ_t represents the model parameters at step t, η is the learning rate, and $\nabla_\theta L(\theta_t)$ is the gradient of the loss function concerning θ_t.

- **Learning Rate**:

The step size, or learning rate η, is critical in determining how fast or slow the model learns. A small η leads to slow convergence, while a large η can cause the model to overshoot minima.

- **Adam Optimizer**: The Adam optimizer improves upon SGD by incorporating adaptive learning rates and momentum, which smooths out updates and accelerates convergence.

The update rule for the Adam optimizer is given by

$$m_t = \beta_1 m_{t-1} + (1 - \beta_1) \nabla_\theta L(\theta_t)$$
$$v_t = \beta_2 v_{t-1} + (1 - \beta_2)(\nabla_\theta L(\theta_t))^2$$
$$\hat{m}_t = \frac{m_t}{1 - \beta_1^t}$$
$$\hat{v}_t = \frac{v_t}{1 - \beta_2^t}$$
$$\theta_{t+1} = \theta_t - \frac{\eta \hat{m}_t}{\sqrt{\hat{v}_t} + \epsilon},$$

where m_t and v_t are the biased-corrected estimates of the first and second moments of the gradients, respectively, β_1 and β_2 are the decay rates for these estimates, and ϵ is a small constant to prevent division by zero.

6.13.4 Techniques to Stabilize GAN Training

- **Gradient Penalty**: A common technique to stabilize GAN training, especially in Wasserstein GANs (WGANs), is to add a gradient penalty term to the loss function [19]. This penalty ensures that the gradients of the critic's output concerning its input are close to 1, preventing vanishing or exploding gradients.

The modified loss function can be expressed as

$$L_{GP} = L_{WGAN} + \lambda \cdot \mathbb{E}_{\hat{x}}\left[\left(\|\nabla_{\hat{x}} D(\hat{x})\|_2 - 1\right)^2\right],$$

where λ is a regularization parameter, $\hat{x}$ is a random sample from the data distribution, and $D(\hat{x})$ is the discriminator's output for the sampled input.
- **Batch Normalization**: Batch normalization is another technique that stabilizes training by normalizing inputs to each layer, reducing internal covariate shift and allowing for higher learning rates.

The normalization can be represented as

$$\hat{x} = \frac{x - \mu}{\sigma + \epsilon},$$

where μ is the batch mean, σ is the batch standard deviation, and ϵ is a small constant added for numerical stability.

6.14 Identify the Benefits and Applications of Generative Models

The most popular generative models are GANs, VAEs, and recently proposed Diffusion Models that have emerged as game changers in several industries due to their ability to generate new data. They are involved in many fields such as art, design, and entertainment among many others and have so many uses and applications.

6.14.1 Advantages of Generative Models

- **Creative Flexibility**: The generative models are a great advantage since they are unique and can help artists and designers create new and creative pieces. Such models are capable of producing numerous versions of images, music, or other content and they can be truly creative since they do not rely on the present dataset.
- **Content Generation**: The models can independently create abstract work—images, music, or virtual world itself. This capability minimizes the efforts needed in the creation of new products enhancing the chances of refining special outputs.
- **Customization and Personalization**: It means that when one trains the generative models in a specific type of input such as text description or a user's preference, one can generate outputs according to the preference. This aspect is very useful where the products are unique or personalized such as fashion designing, production of customized goods and services, movies, shows, music, and the like.
- **Adaptability**: These models can also accept several types of input and hence the ability to produce outputs that meet certain specifications. This adaptability is especially important

for industries of goods that are often customization and personalization important for satisfying the consumers.
- **Efficiency in Content Creation**: By the way, generative models contribute a lot toward simplification of the content generation process. They can create numerous drafts of a design or an artwork which makes it possible for the author to vary works in a short time. The fact that this is possible at such speed is especially advantageous in areas in which speed and innovation are imperative.
- **Rapid Prototyping**: They are useful in helping to design technical solutions more quickly and at the same time come up with several models and then choose from among those that work best.

6.14.2 Applications of Generative Models

- **Art and Design**: There are several applications of generative models, and most of the applications can be linked to art and design. These models are used by artists to develop digital artwork and trying out different themes in the artistic realm. It is quite common for designers to employ generative models to come up with unique designs ranging from clothing and accessories to furniture and structures.

    – AI-Generated Art: With the help of GANs, artists can then come up with new unique pieces of Digital Art based on something abstract or they can invent a new entirely style of art that cannot be achieved manually.
    – Product Design: In industrial designs, generative models help in the generation of different layout configurations with regard to product usability and physical appearance.

- **Entertainment**: In entertainment, generative models are used in the creation of various products such as music, video games, and virtual environments among others. These models help in the production of game levels, musical scores even full virtual worlds for the user thereby improving the user experience.

    – Music Generation: VAEs and GANs can be used to generate new songs which can be in different genres and styles to be used to compose new scores for movies, games, and other productions.
    – Game Development: In video game development, generative models help to build the procedurally generated content like levels, characters, and environments which offer a different set of experiences each time for the players.

- **Fashion and Retail**: The use of generative models proves useful in the fashion industry, especially in designing clothing and accessories. These models can design new fashion styles, "feel and touch" fabric surfaces, and "individualize" designs to the consumers' preferences.

    - Virtual Try-Ons: They facilitate the creation of virtual fitting where the customer is able to envision how a particular piece of clothing would look like on them without having to purchase the clothes. It is therefore a complementary technology that improves the online shopping experience.

- **Healthcare**: In the medical context, generative models can be applied to drug discovery, medical imaging, and in generation of synthetic medical data. When applied in the relevant contexts these apps can increase the pace of knowledge discovery and enhance diagnostic success.

    - Drug Discovery: Generative models aid in making drug discovery by mimicking chemical entities and estimating the probable result an element may have on a biological target.
    - Medical Imaging: The use of GANs is in improving the quality of medical images, for instance, MRI scans by either generating higher resolution pictures or filling in missing information.

- **Marketing and Advertising**: It uses generative models which, for example, can produce individualized advertisements, product usage demonstrations, and marketing material right for a particular niche. This capability assists businesses in better focusing on their marketing initiatives.

    - Customized Campaigns: By employing generative models in advertising, businesses can develop strategies that will appeal to specific audiences and will yield higher results from the audience.

6.15 Related Work

Here are some papers and their contributions to the field:

1. **A—Analyzing StyleGAN3: Alias-Free Generative Adversarial Networks**
   Karras et al. [20]

In this paper, the challenges associated with aliasing artifacts in high-resolution image synthesis using GANs are presented. Karras et al. introduce StyleGAN3, an improved version of the StyleGAN architecture that addresses the aliasing issue. Through experiments, the authors demonstrate that StyleGAN3 exhibits better image quality, and the quality is more coherent thanks to new architectural changes and training approaches. They also provide experimental results showing that, with the help of the alias-free design, the model can produce higher quality images at all scales while decreasing artifacts during image synthesis.

2. **Denoising Diffusion Probabilistic Models** Nichol and Dhariwal [21]

   In this paper, Nichol and Dhariwal utilize Denoising Diffusion Probabilistic Models (DDPMs) and capture several improvements, particularly in enhancing the quality of images generated by the models. To provide readers with a clear understanding of their concepts, the authors present various enhancements, such as improvements to the network structure, the training process, and sampling methods. These enhancements enable DDPMs to achieve higher quality image generation in terms of fidelity than prior work, with fewer artifacts in the images. The paper also broadens the discussion on modifications that can be implemented in the loss function and noise schedules, which assist the model in navigating the diffusion process. Consequently, these developments extend the capabilities of diffusion-based image synthesis, placing DDPMs on par with other generative models like GANs.

6.16 Summary

A new revolution is underway in the field of image creation, especially diffusion models and other text-to-image generation AI tools [22]. Less-known diffusion models work according to the natural process with a peculiar image generation process, where the data is first fed with noise and then reconstructed. All or any of these models help strike the right balance between somehow simplifying things while at the same time retaining many details when used alone or even incorporated in the latent space of autoencoders [23]. New cross-attention layers increase the efficiency of diffusion models in dealing with diverse conditional input thus generating high-fidelity and semantically accurate images.

The text-to-image generation tools include DALL-E 2, Stable Diffusion, and Midjourney, among others with each tool applying unique methods of translating a text description into a colorful image. DALL-E 2 has achieved state-of-the-art image quality, which is good enough for professional work, while Stable Diffusion provides an easier way to generate images, and Midjourney emphasizes aesthetics. The above-mentioned tools not only relate to both text and image media but also offer new opportunities in different fields.

Some of the areas where their applications can be seen range from content production and design to architecture, entertainment, and education, among others. In the future, due to the development of the technology, diffusion models and the text-to-image generators are expected to transform ideas, designs, and messages. They would be able to combine language with images alter industries, improve upon user experiences, and introduce fresh approaches to artistry. As technology advances and the areas of application increase, AI-generated image generation is a promising industry for constant development and creativity.

6.17 Multiple-choice Questions

In this section, you'll find a series of multiple-choice questions designed to test your understanding of key concepts in generative AI. Choose the correct answer for each question.

1. Which of the following correctly describes the purpose of the Evidence Lower Bound (ELBO) in Variational Autoencoders (VAEs)?

   A. To directly compute the exact likelihood of the observed data from the latent variables.
   B. To maximize the Kullback-Leibler divergence between the approximate posterior and the prior distribution.
   C. To approximate the intractable log-likelihood of the observed data by providing a lower bound that can be optimized.
   D. To reconstruct the input data by minimizing the difference between the observed data and the latent variables.

2. In the architecture of a Variational Autoencoder, how does the encoder contribute to the generation of new data samples?

   A. The encoder reconstructs the input data from the latent variables, directly generating new samples.
   B. The encoder maps the input data to a latent space and learns an approximate posterior distribution from which latent variables are sampled.
   C. The encoder applies noise to the input data and outputs a high-dimensional representation for the decoder.
   D. The encoder directly models the likelihood of the data given the latent representation without any sampling.

3. In the context of GANs, which of the following accurately describes the objective function used to train the generator and discriminator?

   A. The generator aims to maximize the discriminator's accuracy in distinguishing real data from generated samples.
   B. The discriminator tries to minimize the generator's ability to create realistic samples while the generator seeks to maximize the likelihood of generating samples that are classified as real.
   C. The generator and discriminator are both trained to maximize the log-likelihood of the real data distribution.

D. The generator's goal is to minimize the log-likelihood of the real data, while the discriminator's goal is to minimize the log-likelihood of generated data.

4. What distinguishes the roles of the generator and discriminator in a GAN architecture?

   A. The generator classifies data as real or generated, while the discriminator generates data samples from random noise.
   B. The generator produces synthetic data samples to resemble real data, while the discriminator evaluates the authenticity of data samples and differentiates between real and generated data.
   C. The generator's primary function is to enhance the resolution of images, while the discriminator performs style transfer tasks.
   D. The generator and discriminator both use convolutional layers to perform data augmentation tasks.

5. In Denoising Diffusion Probabilistic Models (DDPMs), how is the reverse process mathematically represented during training?

   A. By minimizing the noise added in the forward process through a learned score function.
   B. By learning a conditional probability distribution with parameters $\mu_\theta(x_t)$ and $\sigma_\theta(x_t)$ to reconstruct the original image.
   C. By applying a stochastic differential equation to model the diffusion process.
   D. By maximizing the likelihood of observing the data given the diffusion process.

6. What distinguishes Score-Based Diffusion Models (SBMs) from other diffusion model types in terms of their generation process?

   A. SBMs use a score function to measure the likelihood of an image at a given noise level and are trained using adversarial techniques.
   B. SBMs rely on stochastic differential equations to describe the evolution of data through random processes.
   C. SBMs add noise to images using a Gaussian diffusion approach and then reconstruct images from latent representations.
   D. SBMs focus on maximizing the probability of real images in the dataset through a latent representation model.

7. In the DALL-E 2 text-to-image generation process, what role does the "Prior" model play, and how does it contribute to the image generation workflow?

   A. It encodes the textual descriptions into CLIP text embeddings before decoding.
   B. It converts the CLIP text embeddings into CLIP image embeddings, which are then used for final image generation.
   C. It directly generates high-quality images from textual descriptions without the need for CLIP text embeddings.
   D. It refines the noise in the image generation process to match the textual description, effectively generating the final image.

6.17 Multiple-choice Questions

8. Why does DALL-E 2 use a diffusion model as the Prior for converting CLIP text embeddings into CLIP image embeddings?
   A. The diffusion model is more efficient at encoding textual descriptions into high-dimensional vectors compared to CLIP.
   B. The diffusion model is effective in progressively refining noise into coherent image representations, improving image quality.
   C. The diffusion model directly creates the final image from textual descriptions without intermediate embeddings.
   D. The diffusion model simplifies the CLIP training process by eliminating the need for a shared representation space.

9. In the Latent Diffusion Model (LDM) used by Stable Diffusion, how does the diffusion process differ from traditional diffusion models?
   A. It adds noise directly to the pixel data of the image rather than to the latent space.
   B. It operates in the latent space of pre-trained autoencoders, adding noise to compressed representations.
   C. It skips the noise addition step and directly generates images from textual descriptions.
   D. It employs a GAN-based approach to diffusion without using latent representations.

10. What is the primary role of the "Prior" model in DALL-E 2's image generation process?
    A. To encode textual descriptions into high-dimensional vectors for image generation.
    B. To convert CLIP text embeddings into CLIP image embeddings using a diffusion model.
    C. To directly produce images from textual descriptions without intermediate embeddings.
    D. To enhance the CLIP training process by mapping images and texts into a joint space.

11. In Midjourney's use of Conditional Generative Adversarial Networks (cGANs), what is the function of the discriminator?
    A. To generate images from random noise.
    B. To evaluate the quality of images based on random noise inputs.
    C. To differentiate between real images and those generated by the generator, considering the conditioning text.
    D. To optimize the text embedding process used by the generator.

12. How does Stable Diffusion's use of the cross-attention layer benefit the model's image generation capabilities?
    A. It reduces the computational complexity of the diffusion process.
    B. It enables the model to focus on different parts of the input data, improving image resolution and detail.

C. It directly generates images from textual descriptions without latent representations.
D. It replaces the need for pre-trained autoencoders by conditioning directly on pixel data.

13. Which aspect of DALL-E 2's architecture allows it to excel in producing detailed and semantically aligned images?

    A. Its use of Conditional GANs for text-to-image synthesis.
    B. The integration of CLIP for encoding text and images into high-dimensional embeddings.
    C. The application of the diffusion process directly on raw pixel data.
    D. The use of latent space diffusion for computational efficiency.

14. What advantage does Stable Diffusion's Latent Diffusion Model (LDM) offer in terms of computational efficiency?

    A. It avoids the use of autoencoders, processing data directly in pixel space.
    B. It applies the diffusion process in a compressed latent space, reducing computational resources while maintaining image quality.
    C. It generates images using a pre-trained GAN, avoiding the need for additional training.
    D. It leverages a cross-attention layer to simplify the training of diffusion models.

15. In Midjourney's approach, how does attention mechanism contribute to the image generation process?

    A. It improves the generator's focus on relevant parts of the text description during image synthesis.
    B. It enables the discriminator to differentiate between real and generated images more effectively.
    C. It enhances the generator's ability to produce images from random noise.
    D. It facilitates the training of the autoencoder used in the latent space diffusion.

6.18 Answers

Below are the answers to the multiple-choice questions from the previous section:

1. (C) To approximate the intractable log-likelihood of the observed data by providing a lower bound that can be optimized
2. (B) The encoder maps the input data to a latent space and learns an approximate posterior distribution from which latent variables are sampled

3. (B) The discriminator tries to minimize the generator's ability to create realistic samples while the generator seeks to maximize the likelihood of generating samples that are classified as real
4. (B) The generator produces synthetic data samples to resemble real data, while the discriminator evaluates the authenticity of data samples and differentiates between real and generated data
5. (B) By learning a conditional probability distribution with parameters $\mu_\theta(x_t)$ and $\sigma_\theta(x_t)$ to reconstruct the original image
6. (A) SBMs use a score function to measure the likelihood of an image at a given noise level and are trained using adversarial techniques
7. (B) It converts the CLIP text embeddings into CLIP image embeddings, which are then used for final image generation
8. (B) The diffusion model is effective in progressively refining noise into coherent image representations, improving image quality
9. (B) It operates in the latent space of pre-trained autoencoders, adding noise to compressed representations
10. (B) To convert CLIP text embeddings into CLIP image embeddings using a diffusion model
11. (C) To differentiate between real images and those generated by the generator, considering the conditioning text
12. (B) It enables the model to focus on different parts of the input data, improving image resolution and detail
13. (B) The integration of CLIP for encoding text and images into high-dimensional embeddings
14. (B) It applies the diffusion process in a compressed latent space, reducing computational resources while maintaining image quality
15. (A) It improves the generator's focus on relevant parts of the text description during image synthesis.

References

1. GM Harshvardhan, Mahendra Kumar Gourisaria, Manjusha Pandey, and Siddharth Swarup Rautaray. A comprehensive survey and analysis of generative models in machine learning. *Computer Science Review*, 38:100285, 2020.
2. Hamed Alqahtani, Manolya Kavakli-Thorne, and Gulshan Kumar. Applications of generative adversarial networks (GANS): An updated review. *Archives of Computational Methods in Engineering*, 28:525–552, 2021.
3. Zeeshan Ahmad, Zain ul Abidin Jaffri, Meng Chen, and Shudi Bao. Understanding gans: fundamentals, variants, training challenges, applications, and open problems. *Multimedia Tools and Applications*, pages 1–77, 2024.

4. Maciej Zamorski, Maciej Zięba, Piotr Klukowski, Rafał Nowak, Karol Kurach, Wojciech Stokowiec, and Tomasz Trzciński. Adversarial autoencoders for compact representations of 3d point clouds. *Computer Vision and Image Understanding*, 193:102921, 2020.
5. Hanqun Cao, Cheng Tan, Zhangyang Gao, Yilun Xu, Guangyong Chen, Pheng-Ann Heng, and Stan Z Li. A survey on generative diffusion models. *IEEE Transactions on Knowledge and Data Engineering*, 2024.
6. Alec Radford. Unsupervised representation learning with deep convolutional generative adversarial networks. *arXiv preprint arXiv:1511.06434*, 2015.
7. Sagar Saxena and Mohammad Nayeem Teli. Comparison and analysis of image-to-image generative adversarial networks: A survey, 2022.
8. Towards Data Science. Diffusion models made easy. *Towards Data Science*, 2024. Accessed: 2024-12-12.
9. Wei Ren Tan, Chee Seng Chan, Hernán E Aguirre, and Kiyoshi Tanaka. Artgan: Artwork synthesis with conditional categorical gans. In *2017 IEEE International Conference on Image Processing (ICIP)*, pages 3760–3764. IEEE, 2017.
10. TensorFlow. Tensorflow dcgan tutorial, 2024. Accessed: 2024-8-12.
11. Tim Salimans, Ian Goodfellow, Wojciech Zaremba, Vicki Cheung, Alec Radford, and Xi Chen. Improved techniques for training gans. *Advances in neural information processing systems*, 29, 2016.
12. Tero Karras, Samuli Laine, Miika Aittala, Janne Hellsten, Jaakko Lehtinen, and Timo Aila. Analyzing and improving the image quality of stylegan. In *Proceedings of the IEEE/CVF conference on computer vision and pattern recognition*, pages 8110–8119, 2020.
13. Christian Ledig, Lucas Theis, Ferenc Huszár, Jose Caballero, Andrew Cunningham, Alejandro Acosta, Andrew Aitken, Alykhan Tejani, Johannes Totz, Zehan Wang, et al. Photo-realistic single image super-resolution using a generative adversarial network. In *Proceedings of the IEEE conference on computer vision and pattern recognition*, pages 4681–4690, 2017.
14. Xintao Wang, Ke Yu, Shixiang Wu, Jinjin Gu, Yihao Liu, Chao Dong, Yu Qiao, and Chen Change Loy. Esrgan: Enhanced super-resolution generative adversarial networks. In *Proceedings of the European conference on computer vision (ECCV) workshops*, pages 0–0, 2018.
15. Wikimedia Commons contributors. Temperature-dependent diffusion, 2023.
16. Robin Rombach, Andreas Blattmann, Dominik Lorenz, Patrick Esser, and Björn Ommer. High-resolution image synthesis with latent diffusion models, 2022.
17. Aditya Ramesh, Prafulla Dhariwal, Alex Nichol, Casey Chu, and Mark Chen. Hierarchical text-conditional image generation with clip latents. *arXiv preprint arXiv:2204.06125*, 1(2):3, 2022.
18. Nikhil Verma. Diffusion idea exploration for art generation, 2023.
19. Susan Athey, Guido W Imbens, Jonas Metzger, and Evan Munro. Using wasserstein generative adversarial networks for the design of monte carlo simulations. *Journal of Econometrics*, page 105076, 2021.
20. Tero Karras, Timo Aila, Samuli Laine, and Jaakko Lehtinen. Alias-free generative adversarial networks. In *Advances in Neural Information Processing Systems*, volume 34, pages 852–863. NeurIPS, 2021.
21. Alex Nichol and Prafulla Dhariwal. Improved denoising diffusion probabilistic models. *arXiv preprint arXiv:2102.09672*, 2021.
22. Sarah K Alhabeeb and Amal A Al-Shargabi. Text-to-image synthesis with generative models: Methods, datasets, performance metrics, challenges, and future direction. *IEEE Access*, 2024.
23. Walter Hugo Lopez Pinaya, Sandra Vieira, Rafael Garcia-Dias, and Andrea Mechelli. Autoencoders. In *Machine learning*, pages 193–208. Elsevier, 2020.

Setting Up the Environment and Implementing LLMs 7

By the end of this chapter, you will:

Understand the practical part of setting up the environment with Python to work with Large Language Models (LLMs) and see how text generation with these models can be done. For this, we will use Hugging Face Transformers, which is an open-source framework that supports pre-trained models such as GPT-2, GPT-3, and others.

7.1 Environment Setup

Specifically, a proper environment should be provided for LLMs to carry out their work before they are implemented. The section here will help you to set up a Python environment and install the required packages.

1. **Step 1** Create a Python Virtual Environment (Optional but Recommended):
   Virtual environment is an isolated environment in Python that enables the user to handle the dependencies of a project without influencing the general Python installation.

```
# Create a virtual environment named 'llm_env'
python3 -m venv llm_env

# Activate the virtual environment
# On Windows:
.\llm_env\Scripts\activate
# On macOS/Linux:
source llm_env/bin/activate
```

2. **Step 2** Install the required family of packages:
After loading the necessary environment, you then have to install other packages needed in Python with pip. As for the dependencies, we will require a torch for the PyTorch framework and transformers from Hugging Face.

```
# Install the necessary packages
pip install torch transformers
```

This command installs PyTorch which is used in deep learning and Transformers which offers pre-trained models and tokenizers.

7.2 Loading a Pre-Trained LLM

The next step after setting up the environment is to transform them to fit a given pre-trained LLM. For illustrative purposes, we will be using a model that has gained a lot of attention, GPT-2 by OpenAI. GPT-2 is a deep learning generative model that has the capability of creating human-like text from the input it is trained with.

1. **Step 1** Import the Required Modules:
Sometimes while working in Python we need to make or create several modules that would aid us in our work.

   First, we are going to need certain classes from the Transformers library, so, let's import them.

   ```
   from transformers import GPT2LMHeadModel, GPT2Tokenizer
   ```

2. **Step 2** Load the pre-trained model and tokenizer again since we will be using it frequently in our code.
Following that, we need to load in the GPT-2 model and tokenizer [1]. The tokenizer along with the input parses the given text and splits it into tokens or features, which are the units for the model, and the model makes its predictions based on these tokens.

   ```
   # Load the pre-trained GPT-2 model and tokenizer
   model_name = "gpt2"
   model = GPT2LMHeadModel.from_pretrained(model_name)
   tokenizer = GPT2Tokenizer.from_pretrained(model_name)
   ```

7.3 Generation of Text Using GPT-2

Finally, we have the model and tokenizer ready and that makes it possible for us to generate the text depending on the input prompt given. This section will explain the usage of model to fill in the blanks of the sentence or to write any creative content.

1. **Step 1** Define the input prompt:

   It is very essential to comprehend the sort of response that is expected when composing an article because it contributes to the type of article that is being written.

   It starts by setting a scenario on which the model will generate elements based on.

   ```
   # Define the input prompt
   prompt = "In a world where artificial intelligence"
   ```

2. **Step 2** Pre-process the input prompt:

   Before it can be fed into the model the prompt has to go through the tokenization process. This process brings about the transformation of the text into numerical data which is in a form that can be processed by the model.

   ```
   # Tokenize the input text and convert it to a tensor
   inputs = tokenizer(prompt, return_tensors="pt")
   ```

3. **Step 3** Generate text now by entering your text in the box below to continue using the model.

   The method generates a sequence of tokens that continue the input prompt from which it has been generated.

   ```
   # Generate text using the model
   output = model.generate(inputs["input_ids"], max_length=50, num_return_sequences=1)
   ```

   Explanation

   (a) `inputs["input_ids"]`: The input to the model is defined through tokenization of the given prompt.

   (b) `max_length=50`: Conveys the maximum or allowable length of the created sequence.

   (c) `num_return_sequences=1`: This is meant to show that only one sequence should be returned.

4. **Step 4** Decode and print the generated text:

   Last but not least, after generating the token sequence convert it back to the text that human can understand and then print it.

```
# Decode the output to readable text
generated_text = tokenizer.decode(output[0], skip_special_tokens=True)

print(generated_text)
```

Output: In this period of rapid development of artificial intelligence, the difference between man and machine is getting fainter. Self-sustaining robots which are fostered by artificial intelligence in the present day perform intricate operations that were previously performed by the human being [2].

Explanation:

(a) `skip_special_tokens=True`: This helps to exclude such tokens as padding or end-of-a-sequence tokens if they appear in the generated text.

7.4 Fine-Tuning the Model

While pre-trained models like GPT-2 provide a solid foundation for many tasks, there are situations where fine-tuning is necessary to adapt the model to specific datasets or tasks [3]. Fine-tuning involves additional training on the pre-trained model using a domain-specific dataset, with only minimal adjustments to the original model's parameters. When preparing the dataset, it is important to justify its size. Begin by assembling the dataset, typically consisting of a collection of text documents that are representative of the domain you wish to model.

7.5 Customizing Text Generation

Several parameters can be adjusted in the generate method of the object to control the text generation process. These parameters allow you to influence the length, complexity, and diversity of the generated text.

Example with Custom Parameters:

```
output = model.generate(
    inputs["input_ids"],
    max_length=100,          # Generate a longer text sequence
    num_return_sequences=,    # Generate multiple sequences
    temperature=,            # Control the randomness of predictions (lower is more de
    top_p=                   # Use nucleus sampling for more coherent text
)

# Print each generated sequence
for i, seq in enumerate(output):
    print(f"Generated Text {i+1}:\n{tokenizer.decode(seq, skip_special_tokens=True)}\n")
```

7.5 Customizing Text Generation

Expected Output: Generated Text 1
Looking at the current trends in the development of artificial intelligence one can see that more and more human qualities are attributed to machines. Robots that work on artificial intelligence algorithms nowadays perform functions that were previously carried out by human beings.

```python
from datasets import load_dataset

# Load a dataset for fine-tuning (for example, the IMDB movie reviews dataset)
dataset = load_dataset("imdb", split="train")

# Prepare the data for training
def tokenize_function(examples):
    return tokenizer(examples["text"], padding="max_length", truncation=True)

tokenized_datasets = dataset.map(tokenize_function, batched=True)
```

1. **Step 1** Fine-tune the model:
   In fine-tuning, it is necessary to create a training loop in which the model is trained with the help of the given dataset. It emerged that the use of Hugging Face Trainer API makes this process easier.

```python
from transformers import Trainer, TrainingArguments

# Define training arguments
training_args = TrainingArguments(
    output_dir="./results",
    num_train_epochs=3,
    per_device_train_batch_size=4,
    save_steps=10_000,
    save_total_limit=2,
)

# Initialize the Trainer
trainer = Trainer(
    model=model,
    args=training_args,
    train_dataset=tokenized_datasets
)

# Start fine-tuning
trainer.train()
```

2. **Step 2** Save and reload the fine-tuned model:
   It is recommended that after modifying a model, the original model should be stored somewhere and then loaded again when fine-tuning is needed.

   ```
   # Save the fine-tuned model
   model.save_pretrained("./gpt2-finetuned")
   tokenizer.save_pretrained("./gpt2-finetuned")

   # Reload the model
   model = GPT2LMHeadModel.from_pretrained("./gpt2-finetuned")
   tokenizer = GPT2Tokenizer.from_pretrained("./gpt2-finetuned")
   ```

7.6 Summary

Incorporating large language models such as the GPT-2 one has to create a favorable environment, load pre-trained models, and then produce the desired text from the input [4]. As such, the developers can further modify and adjust such models to apply LLMs for various usages including content generation and far more focused tasks within different niches. The accurate procedures and the examples of code from this chapter should help to get the base for working with LLMs in projects.

7.7 Multiple-choice Questions

In this section, you'll find a series of multiple-choice questions designed to test your understanding of key concepts in generative AI. Choose the correct answer for each question.

1. **Question 1:**
   What is the primary purpose of creating a Python virtual environment when working with LLMs?

   (a) To isolate project dependencies
   (b) To speed up the training process
   (c) To reduce the model size
   (d) To enhance model accuracy

2. **Question 2:**
   Which library is primarily used for working with pre-trained large language models in this chapter?

   (a) TensorFlow
   (b) Scikit-learn

7.7 Multiple-choice Questions

    (c) Hugging face transformers

    (d) Keras

3. **Question 3:**
   Which of the following commands is used to install the necessary packages for working with LLMs in this chapter?

   (a) `pip install llm_env`
   (b) `pip install torch transformers`
   (c) `pip install gpt2 models`
   (d) `pip install ai-tools`

4. **Question 4:**
   What is the purpose of the `GPT2Tokenizer` class in the Transformers library?

   (a) To generate text from a model
   (b) To evaluate model performance
   (c) To convert text into tokens and vice versa
   (d) To train a new model

5. **Question 5:**
   What does the `max_length` parameter control in the `generate` method?

   (a) The number of words in the prompt
   (b) The maximum number of tokens in the generated text
   (c) The size of the model
   (d) The speed of text generation

6. **Question 6:**
   Which of the following options describes the `temperature` parameter's role in text generation?

   (a) It adjusts the model's learning rate
   (b) It controls the length of the generated text
   (c) It manages the randomness of predictions, with lower values making the output more deterministic
   (d) It selects the best model version

7. **Question 7:**
   What does `skip_special_tokens=True` do when decoding generated text?

(a) It adds special characters to the output
(b) It ignores padding and other special tokens in the output
(c) It speeds up the text generation process
(d) It enhances the model's accuracy

8. **Question 8:**
   Which method would you use to fine-tune a pre-trained model on a new dataset?

   (a) `train_model()`
   (b) `generate_text()`
   (c) `Trainer.train()`
   (d) `fine_tune_model()`

9. **Question 9:**
   How do you save a fine-tuned model for later use in the Hugging Face Transformers library?

   (a) `model.save_transformer()`
   (b) `model.save()`
   (c) `model.save_pretrained()`
   (d) `model.store()`

10. **Question 10:**
    What is the advantage of using the Hugging Face `Trainer` API for fine-tuning a model?

    (a) It automates the process of training a model
    (b) It reduces the model size
    (c) It increases the tokenization speed
    (d) It automatically generates text based on the dataset

7.8 Answers

Below are the answers to the multiple-choice questions from the previous section:

1. (A) To isolate project dependencies
2. (C) Hugging face transformers
3. (B) `pip install torch transformers`
4. (C) To convert text into tokens and vice versa

5. (B) The maximum number of tokens in the generated text
6. (C) It manages the randomness of predictions, with lower values making the output more deterministic
7. (B) It ignores padding and other special tokens in the output
8. (C) `Trainer.train()`
9. (C) `model.save_pretrained()`
10. (A) It automates the process of training a model

References

1. G Bharathi Mohan, R Prasanna Kumar, Srinivasan Parathasarathy, S Aravind, KB Hanish, and G Pavithria. Text summarization for big data analytics: a comprehensive review of gpt 2 and bert approaches. *Data Analytics for Internet of Things Infrastructure*, pages 247–264, 2023.
2. Rajesh Kumar, Annapurna Maurya, and Abhay Raj. Emerging technological solutions for the management of paper mill wastewater: treatment, nutrient recovery and fourth industrial revolution (ir 4.0). *Journal of Water Process Engineering*, 53:103715, 2023.
3. Karolina Gabor-Siatkowska, Marcin Sowański, Rafał Rzatkiewicz, Izabela Stefaniak, Marek Kozłowski, and Artur Janicki. Ai to train ai: Using chatgpt to improve the accuracy of a therapeutic dialogue system. *Electronics*, 12(22):4694, 2023.
4. Gerhard Paaß and Sven Giesselbach. Improving pre-trained language models. In *Foundation Models for Natural Language Processing: Pre-trained Language Models Integrating Media*, pages 79–159. Springer, 2023.

ChatGPT Use Cases

8

By the end of this chapter, you will:

Having observed that we are in the age of Generative AI, ChatGPT is robust and can transform processes across all industries. Starting with a reconsideration of business and customer service, to changing content generation, marketing, and communication, ChatGPT has numerous applications. The use of ChatGPT has become mandatory in industries and other sectors, including software development, healthcare, marketing and research, creative writing, education and training, legal compliance, human resources, data analysis, etc. In this chapter, the study assesses how ChatGPT has been applied across numerous domains while emphasizing its versatility and the enormous impact it has in addressing multiple issues and supporting decision-making processes.

8.1 Business and Customer Service with Chat GPT

8.1.1 Changing the Face of Customer Support

ChatGPT has brought about a shift in the traditional approach to customer support by providing instant continuous support [1]. It can accommodate a large amount of information from as simple as Frequently Asked Questions (FAQ) to sophisticated ones such as problematic troubleshooting questions. Evaluating the program's natural language richness, one can note that ChatGPT provides unique and hardly distinguishable from human-like answers to customers' inquiries, and the responses are given immediately.

- **Example**: Let's consider a client who enters an e-commerce site of a company that he/she has never dealt with and has no idea where it is located. Not having to wait for a human representative, the customer communicates directly with ChatGPT at least at the beginning of the dialog. The AI gets the query, searches for the product in question and gets back with detailed information. This results in a happy customer who has received the necessary information he or she was looking for without any delay.

8.1.2 Enhancing Sales and Providing Product

By increasing information about sales and sharing information concerning products and services is one of the ways through which ChatGPT can be utilized as a virtual salesperson [2]. It helps customers in their decision-making process by providing information regarding specifications, price analysis, and customer-specific suggestions.

- **Example**: An example of an actual user involves a potential buyer who is visiting an online electronics shop to purchase a laptop. The buyer has a clear purpose for buying the product, as in what he or she intends to use it for such considerations as the amount of money to spend, the performance or power of the product, and how easily the product can be carried around. The buyer interacts with ChatGPT, the latter asks the buyer questions on these needs and offers laptops that meet these needs. It even analyzes features of different models to assist the buyer in decision-making toward the right purchase.

8.1.3 Emphasizing the Continuity of Feedback Analysis

Businesses can utilize the ChatGPT to analyze customer feedback received in form of reviews, surveys, or the feedback customers give on their social media platforms. Thus, by analyzing sentiment data and defining trends, ChatGPT finds hidden potential that will help to adapt products and services.

- **Example**: An online survey where restaurant chain gathers comments from customers. By using ChatGPT to analyze this feedback, the management discovers a recurring theme: customers feel that slow serving and poor presentation of food are bad, to the extent of complaining about them. With these concerns in mind, the restaurant can make specific adjustments that deal with the above challenges to satisfy the customers hence leading to increased sales (Fig. 8.1).

8.1 Business and Customer Service with Chat GPT

Fig. 8.1 AI-driven customer support: ChatGPT enhancing business interactions (created by the authors)

8.1.4 Recommending Personalized Products

With the help of data about customers' preferences and their past actions, ChatGPT can recommend specific products to customers. Using details of past communication, websites visited, and previous purchases that the customer made, it can suggest products or services that the same customer could be interested in.

- **Example**: A user is browsing through an online shop of clothes with the intention of finding the latest fashion wear. With the help of several options that have been bought or viewed by the user before, ChatGPT offers the appropriate clothing and accessories in terms of the individual's preferences. Such a strategy does not only enable the customer to achieve what they want but also make the shopping experience enjoyable thus improving their chances of making the purchase.

8.1.5 Real-Time Order Tracking and Status Updates

A frequently asked question by a customer is where their order is. Picking up from where we left off; with Chat GPT, it is possible to get real-time track information including the current status of the shipment, expected time of arrival, and any contingencies that might be ailing the shipment in the process.

- **Example**: A customer who has recently ordered a product would like to inquire how soon he will receive his order. Without any delay, ChatGPT fetches the recent tracking details and tells the customer that the package is en-route and should be delivered before the day ends. This saves the customer a lot of time since he or she does not have to enquire further, and this enhances their confidence in the firm.

8.1.6 Reducing Returns and Refunds

The return and refund management is usually cumbersome to the customers. ChatGPT makes this process easy by allowing the customers to be led through the steps. The return policy section outlines the company's policy on such returns, offers guidelines on how the item should be packed, and, in some cases, the company may offer to help the client create the return label.

- **Example**: A customer buys a certain product and while using it, he realizes that it has some defects he or she would like to return it. This is where ChatGPT intervenes by giving precise information regarding the return policy, advising the customer on how best they can pack the item, and having the return label generated and forwarded to the customer that he/she can attach to the package. This approach is efficient and effective in minimizing difficulties experienced and therefore improving the customers' experience.

In business and customer service contexts, ChatGPT engages customers, improves support, and makes customer interactions unique. While it is effective in addressing most of the inquiries from customers, it sometimes can take longer to respond to all the queries and may require the involvement of a human being such as when delicate issues are involved. Also, there is a need to establish ethics on the side of the business when it comes to using the data and the frequent mention of AI when it comes to interacting with customers.

8.2 Content Creation and Marketing with ChatGPT

8.2.1 Blog Post and Article Generation

ChatGPT is useful for content writers and may be used to write blog posts and articles on several topics. In this case, ChatGPT can be programmed to search for relevant details and come up with informative content that is organized properly. This capability is most valuable

8.2 Content Creation and Marketing with ChatGPT

Fig. 8.2 AI-powered content creation: a travel marketer collaborates with ChatGPT to effortlessly generate a comprehensive Paris travel guide, covering everything from top attractions to culinary delights (created by the authors)

for companies and individuals who have to make constant publication updates or increase their publishing pace significantly.

- **Example**: Let's consider the example of a travel company that has to release destination guides more frequently to maintain the audience's interest. For example, to prepare a new guide, instead of writing it from scratch, the company can utilize ChatGPT to develop holistic guides on sundry destinations. For example, ChatGPT can prepare an elaborate travel guide on Paris, including tourism highlights, food options, cultural aspects, and general travel information. This enables the company to create high-quality content within the shortest period possible (Fig. 8.2).

8.2.2 Social Media Content Creation

Sharing great and timely content on social media platforms is a mammoth task, mostly for organizations that wish to represent a powerful image online. It is possible to use ChatGPT to create and post ideas, captions, and even replies to comments left by users. It also modifies the content according to the format and the general tone of each platform while keeping the brand image in mind.

- **Example**: The goal of a fashion brand is, for example, to share new outfit ideas on Instagram daily. ChatGPT can come in handy by providing engaging captions that not only depict the outfits but also share advice on how to style them. This is not only time-saving but also equally useful to keep the brand active on social media without putting much effort.

8.2.3 SEO-Friendly Content Creation

Search engine optimization, commonly referred to as SEO, is a very important tool in enhancing the visibility of any website search for ways to optimize their websites in Google. It is possible to create content that is SEO-friendly by making the keywords and phrases used flow naturally throughout the ChatGPT. This increases the chances of the content ranking higher on the search engine results page thus increasing the organic traffic to the website.

- **Example**: The owners of a home improvement company plan to write articles on DIY (Do It Yourself) projects to promote one's website. Articles that ChatGPT create could be titled as "How to Build a Bookshelf" or "10 DIY Home Improvement Projects." By using the most sought keywords within the home improvement and creating area, the webpage will attract more visitors.

8.2.4 Communicative Case

The proper email marketing campaign is rather important to effectively reach out to the customers and influence their decision to purchase the products. ChatGPT is also valuable in crafting hook-laden email copy which creates urgency, as well as calling attention to any special offer. It also assists companies in designing effective email content whose goal is to enhance the overall open rates as well as the conversion rates.

- **Example**: An e-commerce business is planning for a one-time sale during a particular season. ChatGPT assists in the following way: It drafted the email campaign which informs the target audience about the sale items effectively and draws attention to the available discounts using phrases like "Shop Now" or "Limited offer."

8.2.5 Product Descriptions

Generating unique and catchy product descriptions of new products being introduced in the market may be quite a tiresome process, especially for firms dealing with numerous products. This is made easier by ChatGPT where one only has to provide the program with the necessary products and it will output descriptions of the products paying special attention to the features and benefits.

- **Example**: A technology merchant firm selling gadgets such as Smartphones is introducing a new product in the Smartphone market. ChatGPT can easily generate short but accurate descriptions of the phone's characteristics, which include camera resolution, battery capacity, and novelties. These descriptions go a long way in ensuring that the potential customer gets convinced to go ahead and get the product (Fig. 8.3).

8.2.6 Consistent Brand Messaging and Tone

The brand language that companies use in their different channels must be consistent to create a coherent personality for the brand. ChatGPT can easily check all written texts to

Fig. 8.3 A futuristic AI assistant brainstorming creative content ideas with a human marketer (created by the authors)

guarantee compliance with the brand's concept and its values, including social network publications, website content, or marketing collateral.

- **Example**: A fitness firm would like its followers on social media platforms to be encouraged and inspired. Exercise ChatGPT provides information that motivates people to start exercising, lead a healthy life, and be consistent with their habits. Whether it is a text to motivate a user to start the day with an exercise or the text of a new product that has been introduced, the text corresponds to the values of the brand and has an appeal to the target audience.

In content creation and content marketing, ChatGPT enhances the efficiency of content creation to allow marketers as well as writers to focus on the formulation of the best strategies while at the same time providing quality content regularly. However, since the AI-generated content is different from the human-generated one, the content has to go through editing to fit the brand's tone. Such an approach is necessary to have human supervision to guarantee that the content of each part corresponds to the brand's vision and will be interesting and relevant for the target consumers.

8.3 Computer Programming and Technologies Assistance with ChatGPT

8.3.1 Code Assistance and Debugging

ChatGPT also specifically can assist developers in real time to solve their coding-related problems [3]. From syntax mistakes to logical errors and performance issues, ChatGPT is capable of scanning through the code, diagnosing the problem, and recommending the best solution on the matter. This can help avoid time wastage and also help the person learn from his/her mistakes.

- **Example**: For example, a developer, who is typing the Python script, has faced SyntaxError. The error message may be obscure or the developer himself/ herself may not mentally visualize what has gone wrong. When the code has been given to ChatGPT, the tool can automatically point out that the developer erred by failing to use a colon for a loop statement and correct it.

8.3.2 Explanation of Technical Concepts

Algorithms, data structures, system design principles, or any other technical subject for that matter are challenging. With the help of ChatGPT, complex ideas are translated into an easily comprehensible language, in this way, learning is facilitated. Regardless high school students or middle school students, employees, or employers, everyone can utilize ChatGPT to gain more insight into a particular area of study.

- **Example**: For instance, let us assume a computer science student who seems to separate himself from others because he has a lot of questions about recursion. ChatGPT might explain recursion as a process where a function makes a call to itself with a smaller part of the problem. Moreover, to make the student excited, the program can give an analogy that the student would understand easily such as breaking the large problem into several sub-problems that are all similar in solution formulation.

8.3.3 Tech Troubleshooting and Problem-Solving

In complex situations, it can be quite challenging to identify what is wrong with the technical aspect you are using if you have no idea how to get started with it. ChatGPT is capable of a series of questions, from questions about the difficulty to analyzing the reason behind the problem and solutions that can be faced step by step. This can be as a result of a hardware malfunction or as simple as software bugs.

- **Example**: Suppose a user's printer suddenly stops working. ChatGPT might ask questions like, "Is the printer connected to the network?" or "Are there any error messages displayed?" Based on the responses, ChatGPT could then suggest checking the printer's connection settings, ensuring there's paper in the tray, or reinstalling the printer drivers (Fig. 8.4).

8.3.4 Learning New Programming Languages

Transitioning to a new programming language can be overwhelming, especially when you're accustomed to different syntax and paradigms. ChatGPT can ease this transition by providing tailored examples, explaining key differences, and offering resources for further learning. This guidance helps developers quickly get up to speed with the new language.

Fig. 8.4 A user is troubleshooting their printer issue with the help of ChatGPT (created by the authors)

- **Example**: Some considering the transition from using Python to JavaScript might find it hard to deal with the event-driven system in JavaScript. ChatGPT might give an example of how to write a simple function in JavaScript and point out the difference in syntax for function declaration and variables comparison to Python as well as explain what callback is which is not very frequently used in Python.

8.3.5 Documentation and API Usage

APIs—Application Programming Interfaces—are fundamental for today's software, but the question is what about their documentation? Documentation is made easier by ChatGPT; it will also guide you on how to use particular API functions and will also write an example of how to use APIs for your projects.

- **Example**: A developer wants to extend an existing Web application, which is an online store, with a payment option through a third-party payment option. I can introduce them to the API documentation where I explain to them how to authenticate, initiate a transaction, and handling of errors. It can also include an example of the code snippet of a basic API call, for instance, making a payment processing.

8.3.6 Software Best Practices

It is important to write clean code that can be easily maintained and that is rather effective in the long-term perspective. It is easy to explain that ChatGPT can explain to developers various tips, tricks, patterns, and principles regarding coding styles and generating good quality software. It is quite helpful, especially for junior developers or any developer who wants to improve his or her skills.

- **Example**: For example, a junior developer who was assigned their first large-scale project may come asking how they need to structure their code. ChatGPT could recommend good practices such as dividing the code into modules (modular programming), writing comments to explain the code's complex logic, or incorporating the use of a version control system such as GitHub for collaboration.

As a tool in software development and technical support, ChatGPT helps reduce development cycles and improve instruction and issue-solving. Nonetheless, developers should use their personal best judgment especially when situations are a bit complicated most of the time ChatGPT's suggestions do not consider particular circumstances.

8.4 Data Entry and Analysis with ChatGPT

The transformation of ChatGPT into the code interpreter, now known as the "Advanced Data Analysis" (ADA) tool, represents a significant leap in its capabilities. This evolution has equipped ChatGPT with enhanced features that make it a powerful ally for data professionals. Beyond just coding assistance, ChatGPT now offers expertise in advanced data analysis techniques, statistical modeling, data visualization, and more. This expanded functionality allows users to derive deeper insights from their data, making ChatGPT an indispensable tool for a wide range of data-driven tasks. Let's explore how ChatGPT supports data entry and analysis.

8.4.1 Data Entry Assistance

Data entry can be a time-consuming and error-prone task. ChatGPT streamlines this process by accurately transcribing handwritten or typed content into digital formats, inputting information into spreadsheets or databases, and organizing data according to predefined formats or categories. This not only saves time but also reduces the likelihood of human error.

- **Example**: A research team is conducting a large-scale study that involves collecting responses from paper surveys. Rather than manually entering each response, which could take days, the team uses ChatGPT to transcribe the handwritten responses into a digital spreadsheet. This tool efficiently handles the data entry, ensuring that the information is accurately and quickly digitized, allowing the team to focus on analysis.

8.4.2 Data Cleaning and Preprocessing

Data Entry and Analysis including ChatGPT

The change of the ChatGPT into the code interpreter, which is now called the "Advanced Data Analysis" or the ADA tool, is a leap forward. This evolution has endowed ChatGPT with better features that in turn make it a worthy tool for data professionals. It is no longer a simple coding help tool, now, through prompting ChatGPT questions, users can get answers, code samples, or even guides involving educational subjects such as DA, statistical modeling, data visualization, and much more. These new features enable users to gain more insights about their data making ChatGPT essential in various data processing activities. Let's explore how ChatGPT supports data entry and analysis.

- Data Entry Assistance: The process of data entry involves keying vast volumes of information to a computer which may be boring and requires a lot of time as well as being prone to many mistakes. ChatGPT helps in this respect by typing written or printed text and converting it to digital values, entering text in tablets or spreadsheets, as well as categorizing texts according to format or pre-selected categories. This not only saves time but also minimizes the chances of having wrong data input by human beings.

  - **Example**: A group of researchers has been designed a massive-scale research that requires obtaining data from paper questionnaires. The team does not type each response themselves because it would take days; instead, the Typed team uses ChatGPT to translate the handwritten responses into a digital Google Doc. This tool effectively manages the input of data into the system thus minimizing the time, energy, and effort used in data entry and hence the team gets to focus on the analysis.

- Data Cleaning and Preprocessing:

The first thing that has to be done before any proper analysis can be made is to make sure the data collected are clean and properly formatted. Many consider data cleaning and data preprocessing as important steps in this process. Some of the actions that you can perform with ChatGPT include issues such as normalization, which is the process of looking at the data for the presence of inconsistencies, and where found, defining them and correcting them; other activities include standardization whereby if a value is missing

8.4 Data Entry and Analysis with ChatGPT

in the set or if there are some values which are not within an expected range, ChatGPT flags such values. This helps to ensure that the data is in the right condition for analysis which in turn helps to improve the results obtained.

– **Example**: In the process of preprocessing a dataset for machine learning purposes, an analyst discovers that some data are duplicated while others are missing. It is possible to inform these issues in ChatGPT and the latter will be able to identify these issues, indicate entries that can cause problems, and even provide ways to solve these problems, for example, to delete similar entries or complete missing data with mean or median values. This process of cleaning the data automatically makes the dataset to be fit for accurate and correct analysis.

8.4.3 Basic Data Analysis and Visualization

When it comes to the calculation and generation of simple statistics (mean, median, standard deviation, etc.), generation of charts, and summary of trends, a simple analysis can be quickly completed using ChatGPT. This capability enables users to get summary results and make decisions depending on such results (Fig. 8.5).

Fig. 8.5 Marketing team analyzing sales data for strategic planning (created by the authors)

8.4.4 Data Interpretation and Insights

Knowledge is more than simply the interpretation of statistics; it's comprehension. ChatGPT helps in analysis by doubling double of data and correlating the data with various patterns that are explained in detail and about important findings. This assists the users in making sense of their data and incorporating it into decision-making.

- **Example**: A web analyst looks at the number of visitors and realizes that it has decreased significantly in a certain period. ChatGPT recalls the information and comes up with potential reasons which might be a change in the search engine algorithm, server issues, or even fluctuations in user habits due to time of year. In so doing, ChatGPT assists the analyst in identifying the main reason for the traffic drop and remedies the situation.

8.4.5 Comparative Analysis

Pairing up both datasets or within the same set of variables is extremely important in observing the relation, difference, or similarity between the datasets. To this end, ChatGPT supports comparisons and assists users in making appropriate conclusions from the data they have.

- **Example**: A business needs to get the customer satisfaction rate for two of its products in order to make a comparison. ChatGPT computes the mean of satisfaction scores for each product line, compares and contrasts it, and gives an account of what could be plausible for the difference observed. This information helps the business to determine which product line is more popular in regard to the satisfaction level and to what extent.

8.4.6 Data Reporting and Summarization

This applies to descriptions of data findings since some of the stakeholders might be unable to perform an analysis of data findings on their own. This is made easier through ChatGPT which provides summarized reports complete with findings and trends from the data analyzed. This increases the ease with which the decision-makers will comprehend the data as well as take necessary actions.

- **Example**: An analyst is to provide a quarterly report on sales to the top management of the company. ChatGPT produces reports on the general performance indicators including revenue patterns, sales volumes, and differences per region. The logic of structuring the report, which divides it into sections, makes it easy for the executives to make business decisions based on the sales performance of the company.

ChatGPT helps in a simple way in data entry and analysis where it comes in handy when doing basic analysis and data entry. Yet simple analysis with extensive statistical analysis, measurement, and more subjective interpretation requires human input to draw more accurate conclusions necessary for conducting decisions.

8.5 ChatGPT's Role in Healthcare and Medical Information

ChatGPT can be useful in giving people an accurate and clear explanation or advice concerning different health issues. It is also worth mentioning that despite the information that can be obtained from ChatGPT it's not a healthcare professional advice. Here's a more detailed breakdown of how ChatGPT can assist in healthcare.

8.5.1 General Medical Information

ChatGPT can give general information about the symptoms, diseases, therapies, and ways to protect one's health. Since the information about most of the general health issues is provided, it can assist people and provide the essential knowledge about their health before consulting a professional.

- **Example**: A person suffering from a headache may one day wake up and ask ChatGPT questions like the following: ChatGPT could recommend some advice to avoid frequent headaches and such causes could be stress, lack of water intake, or migraine. It can also give advice on how the discomfort can be dealt with while at the same time stressing the need for one to seek medical attention to be evaluated and prescribed accordingly.

8.5.2 Symptom Checker and Self-Assessment

In this way, for example, ChatGPT asks the user-specific questions and provides the user with a list of possible causes of the described symptoms. It can also indicate when one may need an emergency or probably a regular check-up (Fig. 8.6).

- **Example**: Whereas if a person says that he/she is feeling sick and has a fever, body pains, and general body weakness, ChatGPT may advise the person to seek medical consultation for flu or common cold. It may also advise the patient to take a break and drink lots of

Fig. 8.6 General medical information with ChatGPT (created by the authors)

fluids, observe the symptoms, and see the doctor in case the symptoms worsen or do not clear up in a given duration.

8.5.3 Medication and Treatment Information

With ChatGPT, users can get over-the-counter medication information such as how the medication works, its side effects, and any interaction with other medicines. This can assist users in comprehending their treatment but should be used as a supplement to a conversation with a doctor or a pharmacist.

- **Example**: For example, if a person is put on a new medication, ChatGPT can give information on the general usage of the said medication and a list of possible side effects including but not limited to nausea and/or dizziness. It also advises the user to contact his/her healthcare provider if any adverse or severe reaction occurs.

8.5.4 Wellness Tips and Healthy Habits

Action: ChatGPT presents recommendations to enhance the quality of life and disable diseases. This entails recommending matters related to food, exercise, emotional and mental health, and precautionary measures that can help the users in their day-to-day lives.

- **Example**: Here the user might pose a question like, what can I do to have a better night of sleep? Here, it would be possible to come closer to identifying forms of advice that ChatGPT might provide, for instance, the following factual recommendations regarding teamwork:

  – Establishing a consistent bedtime routine.
  – Reducing screen time an hour before bed.
  – Creating a comfortable and dark sleep environment.
  – Avoiding heavy meals and caffeine late in the day.

  These tips are designed to help users adopt healthier habits for improved physical and mental well-being.

8.5.5 Explanation of Medical Terms

Medical words are sometimes overwhelming and hard to comprehend by people with little or no knowledge of the terms used in the health sector. ChatGPT can explain medical terms, an acronym, or an abbreviation and translate them into simple and comprehensible English.

- **Example**: Suppose a person reads some medical report and comes across the term "hypertension" and they do not know what it means then ChatGPT would tell the person that hypertension is a medical term that means high levels of pressure in the arteries which is the force of blood on the walls of arteries; a condition that is not healthy and can cause health risks if not well managed.

8.5.6 Preparing for Medical Appointments

ChatGPT can be used to engage proactively in a conversation with a user to ensure that they receive the optimal value upon a visit to the doctor by offering questions to ask, tips on how best to explain their symptoms, and other aspects that may be vital to share with the doctor from a patient's health history.

- **Example**: A user who has an appointment next week to talk to a doctor about a persistent illness they have could humbly inquire from ChatGPT questions they need to ask their physician. ChatGPT might suggest:

- "What many interventions could be most effective for the management of my condition?"
- This is especially so if they want to know how they can manage their symptoms daily.
- "Are there some aspects of one's lifestyle that can be changed to improve one's overall health?"

It increases the level of prepared users to attend medical appointments to have full confidence.

8.5.7 Mental Health Support and Stress Management

Mental health is an important aspect of human health and a person can always seek advice from ChatGPT on how to cope with stress, anxiety, and other disorders. ChatGPT is not a therapist but it can give simple ways of coping with the situation, give tips on how to care for oneself, and advise that one should seek professional help.

Example: For instance, a person may request ways of handling stress if he or she is stressed. ChatGPT could suggest:

- Practicing deep breathing exercises.
- Engaging in physical activity like walking or yoga.
- Trying mindfulness or meditation techniques.
- Reaching out to friends, family, or a professional for support.

It's useful for gaining preliminary knowledge and for questions related to health that do not require immediate medical attention. But all the time, users should consult professional healthcare practitioners to get the right diagnosis and treatment advice.

8.6 Market Research and Analysis with ChatGPT

Market research: With the help of ChatGPT, it is easier to work with large amounts of information. Here's a breakdown of how it supports various aspects of market research: Here's a breakdown of how it supports various aspects of market research:

- Survey Analysis and Summarization: Thus, it can be seen that when conducting surveys, businesses may find themselves with an avalanche of data. In this, ChatGPT assists by screening the survey answers and selecting standard topics and trends, for example,

8.6 Market Research and Analysis with ChatGPT

commonly reported issues or novelty. It presents brief notes of the key issues received from the survey and special emphasis is made on either satisfaction or concern.

- **Example**: Suppose a particular company decides to conduct a poll to determine its clients' level of satisfaction. The results can be scanned and filtered by ChatGPT, thus highlighting such aspects as the level of overall customer satisfaction or dissatisfaction; the main issues raised by the respondents.

- Customer Feedback Insights: There are different forms of feedback that customers give to firms which include online reviews, social media comments, and direct comments. ChatGPT categorizes feedback into positive negative and even neutral. It is also useful in identifying shared features or issues, possibly frequently raised by customers.

    - **Example**: When the site is an e-commerce site, ChatGPT can look into product reviews, and the majority if not all of it can be classified as positive feedback, negative feedback, or neutral feedback and it can also give brief of the things that people are saying mostly.

- Competitor Analysis: It's very important to define strategy, and which competitors are. ChatGPT assists competitor analysis by gathering information from different sources about the competitors, their advantages and disadvantages, and plans. What is a detailed and concise overview of competitors' market position and characteristic features?

    - **Example**: A case in the context of a tech startup analyzing competing players in the smartphone space can potentially use ChatGPT to gather all the data about the features, and pricing strategies of the competing players as well as representations of customers where one can see what one is up against.

- Trend Identification and Forecasting: It can therefore be said that keeping abreast with the market trends may make a difference for a business. It supports the analysis of information retrieved from social media activity, news articles, and the latest industry reports for identifying new trends. Used to forecast trends in the future based on the current trends to assist business organizations in making better decisions.

    - **Example**: A fashion brand having a plan on what the next season's popular look will be can use ChatGPT to scour the Internet, specifically social media and fashion bloggers.

- Consumer Behavior Analysis: To market successfully, one has to know why consumers make the choices that they do. ChatGPT helps by analyzing available information on the purchasing pattern, inclination, and driver. Analyze certain patterns about consumers that can help draw up strategies and release new products.

- **Example**: For example, an online retail store that wishes to know why some of its products are popular during certain seasons of the year it can process data of its purchasing to find out the issue.

- Market Segment Profiling: Due to the above analysis, it can be concluded that the activity of effective marketing cannot be imagined without considering various segments of the market. Thus, ChatGPT supports by developing elaborate consumer profiles according to the demographic, geographic, and psychographic criteria. It assists in fine-tuning the strategies that are used in the marketing process in order to target the right markets.

  - **Example**: An electronics manufacturer aspiring to introduce a new product to the market will benefit from ChatGPT in as much as the profiles he creates to appeal to the potential customers will contain detailed descriptions of the customers' preferences and interests.

In this case, ChatGPT helps in improving market research in that it provides quicker results in terms of data analysis and provides a clear decision-making context. However, they base on human skills and experience to assess and apply these findings when establishing sound business strategies grounded on a complete understanding of market trends in the environment.

8.7 They Are Creative Writing and Storytelling with ChatGPT

As the following information, ChatGPT can be a very useful tool when it comes to the creative writing process as an assistant who will provide a certain set of helpful tools to make ideas come to life as well as build a great story out of them. Here's a breakdown of how it supports various aspects of creative writing.

- Idea Generation and Brainstorming: Finder of novel and interesting things is usually the single problem of creativity in writing. ChatGPT can help generate creative concepts for the story, article, or any other project, get that spark, and, perhaps, get rid of writer's block. They can come up with myriad plots, storyline themes and scenarios as well as characters to develop.

- **Example**: If the author of a new novel is blanking she might get the idea of a plot where people in a futuristic society can change their lives which will be a new direction for her (Fig. 8.7).

- Plot Development and Story Outlining: Among the important elements, the narrative's construction plays a big role in the story's coherence as well as the audience's engagement. Sometimes users need assistance in generating such a storyline, and that is where ChatGPT shines because it will assist in the formulation of the main plot, as well as subplots to make the narrative balanced. It is useful in creating suspense and then letting go of it at times effectively keeping the readers hooked. It helps inulation of events in a sequential manner or rather a sequential narrative of the events is easily created.

Fig. 8.7 Creative inspiration: writing a story with ChatGPT (created by the authors)

– **Example**: ChatGPT may serve as a valuable tool for a screenwriter who is developing a TV series pilot; to mention several options, this tool can help outline episodes, introduce characters, and anticipate the further storyline thus maintaining the necessary concept and plot of a show.

- Character Creation and Development: Both the author and Dr. Johnson highlighted the need to have well-developed characters for the smooth running of the story. This is supported by the fact that ChatGPT suggests character attributes, backgrounds, and reasons to the characters. It assists in determining how certain characters will undergo growth in the course of the story.

   – **Example**: A fantasy writer who is working on a new novel's main character may find the ChatGPT tool helpful when it comes to designing a complex hero with a dark background, a hidden power or some other factor that shapes the story.

- Dialog Writing: It specifies a more detailed view of whether the characters in the story have come to life and whether they have started interacting. The role of ChatGPT is to provide ideas of how to continue the conversation and what a character would say in a particular context concerning the character's personality and the general setting of the story. It assists in making dialog that can depict the attitudes of different characters within a certain relation.

   – **Example**: A gearing toward dramatic situation a playwright may likely turn to ChatGPT to write the scripts of the exciting scene that has some hidden conflict.

- Worldbuilding and Setting Descriptions: These environments help in telling the story in a more effective way to help the viewer get into the feel of the story. ChatGPT helps in describing places, the feel of the place, the look of it, sounds and touch felt as one is in that particular place. It also assists in the construction of complicated environments, which possess their laws and features.

   – **Example**: If it is a science fiction story and the story is based on an alien planet, the chat GPT may describe the planet in terms of geographical features, weather conditions, and vegetation so that the author will find it easy to come up with a believable place that the story can take place.

- Creative Prompts and Writing Exercises: There are a number of ways in which flow can be disrupted and one of them is writer's block addressed when engaging in creative prompts and exercises. ChatGPT itself has suggested that it can give stakeholders premises from where one can start short stories, poems, or even other stories. Another section is dedicated to suggestions of exercises that can help one improve various aspects of writing as well as encourage creativity.

- **Example**: A poet in search of a motivation to create a new poem can take ChatGPT's prompt about the light and shadow of nature to overcome the creative blockades.

ChatGPT supports writers in that it provides them with concepts, an outline, and motivation. Nevertheless, I discovered that the writer's voice, judgment, and editing are crucial for creating a captivating and structurally sound narrative. Provence went further and argued that for them generating material was also helpful in the writing process.

8.8 Education and Learning with ChatGPT

ChatGPT is a flexible learning companion that helps the learning process by providing suggestions and materials. Here's a closer look at how it contributes to various aspects of education.

8.8.1 Virtual Tutoring and Concept Explanation

Edit: Some of the features of ChatGPT include answering questions, which makes it a virtual tutor in that it can explain academic concepts and areas of interest in detail. It assists a student in grasping details that are difficult with the help of breaking them down into simple steps and components. It provides examples and cases which guides one through various solution ladders and assists in the solution of exercises and walks through examples and cases to enhance understanding and application of theories.

- **Example**: At the high school level, there could be a case that one is having difficulty making sense of calculus, derivatives, or integrals; this is where ChatGPT comes in, it will explain the concepts and then solve some problems simultaneously.

8.8.2 Homework and Assignment Help

It can be quite overwhelming when one has to write an assignment, but ChatGPT helps by giving guidance on how to write different types of assignments like essays or projects. It is especially useful for answering specific questions and has tips on how to approach many tasks.

- **Example**: Using ChatGPT, a student who is to write an essay on a historical event he or she has been given can easily be assisted to come up with the thesis statement, the main ideas to address in the given topic, and even the format of the given essay as well as its coherence.

8.8.3 Language Learning and Practice

Language learners achieve practice through dialogs with ChatGPT as a tool that assists learners in the use of the language in real-life situations. It provides feedbacks that enable the correction of such skills as well as suggestions for enhancement of the same. It teaches new words and phrases, improving the vocabularies whereas it does it quite naturally.

Example: If a student is studying Hindi, then ChatGPT can reply in Hindi, correct their grammar mistakes, and also note new words within the flow of the conversation (Fig. 8.8).

8.8.4 Study Resource Generation

One of the important processes in getting the whole overall revision is the development of good study aids. ChatGPT is useful in creating flashcards that can be used when there is a need for a quick revision of specific ideas. It can present a great deal of information in the

Fig. 8.8 Bridging languages: learning language with ChatGPT (created by the authors)

form of brief evaluations and recommendations. It offers a section of questions to practice with to check the level of understanding and as well to prepare for tests.

- **Example**: A student preparing for a history exam can say to ChatGPT: "Tell me about important events that I need to revise" or "Help me create practice questions."

8.8.5 Research Assistance

Research is not an easy process, however, ChatGPT can source and provide references for various types of research. It is useful in helping formulate meaningful but specific research questions. It assists in the organization of research work and the presentation of findings and this is because it provides an organized way of presenting information.

- **Example**: A college student, who wants to study climate change, will be able to find sources for his research, formulate certain research questions, as well as get help in outlining the research paper with the help of ChatGPT.

8.8.6 Exploring New Topics and Curiosities

This curiosity is one feature that is fostered with ChatGPT by explaining all sorts of things in simple terms. It states new ideas on the topics, helping the learners to dig deeper.

- **Example**: If a student is interested in quantum physics, he or she can get basic information from ChatGPT which, in turn, would hook such a student with a foundation on which he or she would like to explore more.

Apart from aiding known conventional classrooms, the following are ways through which ChatGPT contributes to learning: conversational learning, encouraging curiosity, and academic assistance. Thus, endangered is the significance of teachers' instructions, certain curriculums, and the very processes of students' tempering of their thoughts.

8.9 Legal and Compliance Support for a Specific Company or Business with ChatGPT

ChatGPT can be of great use in the legal industry, for research, writing, and creation of documents, checking compliance, and more. However, it gives considerable help and it is crucial to underline that the service provides help as a supplement in cases with legal aid. Here's a closer look at how ChatGPT can be applied in legal contexts.

8.9.1 Legal Research and Case Law Analysis

Legal research entails the discovery of the legal authorities such as cases and precedents that support a particular position. This is supported by ChatGPT in that it helps in pulling out relevant details about databases and case laws for easy evaluation by the lawyer. It assists an advocate in locating legal precedents that they can incorporate into their legal argument or as a reference point for other related issues.

- **Example**: A lawyer planning for a case dealing with intellectual property rights might use ChatGPT to get the summary of some previous similar cases that will help him/her in developing the argument because of the existing cases.

8.9.2 Drafting Legal Documents

Preparation of legal instruments entails following the law and putting into consideration the provisions of the law. Some aspects that show that ChatGPT operates in law include

It contains ready-made templates for legal documents, contracts, agreements, and letters among others. It assists in writing documents using correct legal vocabularies, and legal terminologies that may be required. It ensures that the documents provided satisfy legal provisions and legal requisites.

- **Example**: An entrepreneur in a position to require an NDA can use ChatGPT to draft the nondisclosure agreement and assist in putting into the essential working clauses to protect the confidential information and anything related to the creation of the agreement so that the document is legal.

8.9.3 Legal Definitions and Explanations

Resource: Vocabulary related to the legal field may be quite regulating, and confusing for people who are not legal professionals. It makes this process easier by giving simple explanations of legal terms and ideas as ChatGPT does. It provides how certain lawful issues can be implemented in real-life circumstances.

- **Example**: A business owner coming across the term "tort" can ask ChatGPT what the term means. Interested, ChatGPT would explain that a tort is a civil wrong giving rise to legal responsibility and enumerate the various torts and their consequences.

8.9.4 Compliance Guidelines and Regulations

Knowledge of legal standards and the regard of the standards are significant to all entrepreneurial operations. The integration is useful in that chatGPT suggests laws and regulatory compliance specific to industries or practices. The steps to follow point out how a business should make certain that it meets all the legal requirements.

- **Example**: An organization desiring to know more about data protection laws can type its query in ChatGPT and get to know the general guidelines of the laws such as General Data Protection Regulation (GDPR) and advice on how to adhere to these laws.

8.9.5 Preliminary Legal Advice for Common Issues

When it comes to basic legal concerns, ChatGPT helps by giving preliminary advice by outlining the process for legal concerns. It provides the users with beneficial advice and the elementary steps to solve the most common legal actions.

- **Example**: For instance, for a small business owner with no idea of the correct procedure that must be followed in case of employee dismissal, chat with ChatGPT would guide you on measures that must be undertaken regarding legal requirements and policies.

8.9.6 Intellectual Property Guidance

In several ways, getting protection for intellectual property is crucial to the creators of these products and the firms. ChatGPT supports this by giving information about various categories of intellectual property such as copyright, trademark, and patent. The one outlines the procedures that have to be followed in a quest for registering and protecting an IP.

- **Example**: Thinking about an artwork that an artist creates, an individual who wants to protect his work can use ChatGPT to familiarize himself with the rules of the copyright law and the ways to register his creation as well as the measures that can be taken against the infringing.

When used in legal and compliance, ChatGPT helps in increasing effectiveness, smooths the documentation process, and offers opportunities for instant analysis. However, in complicated legal issues and decisions which may have significant legal effects or might affect the legal rights of some people, professional legal advice is imperative.

8.10 ChatGPT for Support to Human Resources and Recruitment Departments

ChatGPT can greatly improve the realization of HR and recruitment, and bring value for a range of activities and operations. Here's a closer look at how ChatGPT contributes to HR functions and recruitment.

- Candidate Screening and Initial Interviews: When dealing with a large number of applicants one may encounter certain difficulties. ChatGPT assists with that purpose by interacting with the candidate through chats and then asking questions related to the educational background, experience, and abilities. It helps to assess the answers and select the individuals that match the requirements of the position and can further proceed to the next level.

    – **Example**: An HR manager, for instance, would be happy to be flooded with resumes for a marketing post. First, ChatGPT interviews the applicants and gets to know their marketing plans and previous accomplishments before selecting the best candidates for the position.

- Job Description Crafting: Coherent job descriptions help in recruiting the best candidates for the job and also communicate the expectations of the job. When creating job descriptions ChatGPT produces accounts of key responsibilities, required skills and qualifications, organizational culture, etc. It makes sure that the requirements of the job definition fit the organizational culture and that the applicants who are applying for the positions also fit into the culture of the organization.

    – **Example**: An organization is in the process of sourcing a new software developer for a specific company. ChatGPT creates a job description that focuses on the technical qualifications of the candidate and the major projects in the organization while showcasing the firm's position that embraces creativity.

8.10 ChatGPT for Support to Human Resources and Recruitment Departments

- Employee Onboarding Support: A good experience for the early days of employment is what promotes the assimilation of the new follower into the organization more effectively. Newcomers at the workplace find information about the company policies, benefits, and procedures from ChatGPT. It responds to the questions concerning the company's vacation policy, medical insurance, as well as onboarding procedure.

    - **Example**: A new employee is hired in the organization and they are not aware of the organization's remote working policy. ChatGPT gives a brief outline of the policy and outlines how one can ask for an arrangement for teleworking.

- Training and Development Assistance: CPL is very important in the growth of the employees and hence the need to support growth continuously. This is facilitated by ChatGPT in that it presents online classes, workshops, and other training that may suit the needs and career paths of the employees. It also allows the employees to learn where they stand in terms of the available programs and where they can learn new things that they need.

    - **Example**: An employee wants, for instance, to acquire leadership skills. Based on the discussion, online leadership courses and workshops that fit the employee's career plan should be listed as per ChatGPT.

- Employee Assistance and Policy Clarification: Appreciation of company policies is very essential in avoiding a situation that may compromise the relations between employees and employers. ChatGPT helps the employees by offering comprehensible information about company rules including leave matters, grievance process, and conduct. It describes how to report a problem or make a complaint and how to make a request.

    - **Example**: An employee is not very sure how to go through the procedure of filing a grievance. ChatGPT also describes how it must be done and who must report it.

- Interview Preparation and Tips: Interviews can be quite daunting for the candidates especially when preparing for the interviews. To avoid ambiguity, the job role of ChatGPT is to assist by giving guidance on how one can go about researching the company or organization, as well as the position held, all the while offering ideas on how to come up with answers for some of the frequently asked interview questions in the job market. It gives guidelines on how one is expected to conduct oneself, especially concerning dress code and language.

    - **Example**: A candidate feels anxious about an interview for a financial job that he or she is to attend. ChatGPT advises on how to answer technical questions, talk about previous experiences, and make the first impression in the interview.

ChatGPT strengthens the human resource and recruitment function in the following ways: it refines the recruitment procedures and optimizes them to benefit both the employer and the employees being recruited; it helps manage the candidate experience in the best way possible; and with its help, the communication processes can be more streamlined. Still, people's participation is necessary for flexibility and critical thinking in the decision-making process, soft skills evaluation, and sophisticated HR situations.

8.11 Personal Assistant and Productivity Using ChatGPT

For instance, ChatGPT can assist with time management and planning, conducting online searches for relevant information, and other tasks that would otherwise be time-consuming. This section demonstrates how ChatGPT addresses productivity-related challenges and highlights tasks that can enhance personal efficiency.

8.11.1 Task Management and Reminders

Staying organized is key to managing tasks effectively. ChatGPT helps users organize tasks into manageable lists, ensuring nothing is overlooked. It can set reminders for appointments, deadlines, and important tasks, helping users stay on track.

- **Example**: A user has planned to have a meeting at 3 PM, and he/she requests ChatGPT to remind him/her at 2:45 PM. In this way, they get a notification that a meeting will be held in advance and this helps them to prepare themselves for it.

8.11.2 Calendar Coordination

Regarding time, it is important to schedule events and coordinate them so that one would not get confused between them. By integrating with users' calendar applications, ChatGPT can suggest appropriate time slots to schedule a meeting or an event. It also assists in availability planning for appointments as well as extension of invites.

- **Example**: We want a professional to schedule a meeting with other workers during a working day, but all of them work in different countries and have different time zones. ChatGPT looks into the calendars of the participants in the meeting and gives the meeting time that everyone is free to attend.

8.11.3 Information Retrieval

Time can be saved and productivity increased if one gets the correct information the first time he or she searches for it. Therefore, ChatGPT assists by searching facts and definitions or statistics and history from the Web or databases. In this case, it provides summaries of complicated issues or bulky reports or issues that one may have little time to peruse.

- **Example**: A student writing a paper on climate change is explaining their situation to ChatGPT which is to help them find modern material in the form of articles and data. ChatGPT locates articles or statistics and sources needed by the student for his/her work.

8.11.4 Note-Taking and Summarization

It is clearer how note-taking as well as the summarization process assists people in documenting as well as reviewing the relevant content. ChatGPT helps by taking notes of the important points that may be discussed in meetings, classes, or a conference. It simplifies complex documents and/or presentations and summarizes them in a few lines maintaining their intent.

- **Example**: Imagine a user in a business meeting invokes ChatGPT and says, report the meeting. The advantage of using ChatGPT is one is able to have a summary of the key points that were discussed during the meeting, as well as the action points as a quick reference.

8.11.5 Language Translation on the Go

Tourists and people who understand many languages gain from getting the translation done in real time. The following are the ways that ChatGPT assists in translating spoken and written words, signs, menus, and other texts from one language to another. It helps in developing skills in speaking and comprehending foreign languages.

- **Example**: Imagine a person is traveling, for example, in Japan and he/she has to translate the menu in a restaurant; therefore, s/he has to turn to ChatGPT and ask for immediate translations of the dishes to be able to make proper choices.

8.11.6 Personalized Recommendations

Information searching and decision-making can be made easier especially when one is presented with a list of items they might want. This way, ChatGPT suggests movies, books, songs, restaurants, and a lot more depending on users' likes and dislikes. It provides information regarding the quality and ratings of suggested items.

- **Example**: A reader who wants to read a new mystery needs a recommendation from ChatGPT. Getting the list of most popular mystery books, ChatGPT offers their brief descriptions and readers' reviews.

8.11.7 Fitness and Wellness Assistance

This can be put into practice by the support of advice concerning physical fitness and general well-being tailored to a person. ChatGPT aids by creating plans for exercise routines depending on the individual's fitness objectives and interests. It has information on weight management, portion control, nutrition, and where to find information on good healthy eating manners.

- **Example**: A user feels the need to begin a home workout regime. ChatGPT customizes an exercise routine, featuring the exercises, sets, and recommended levels of intensity together with advice on nutrition.

ChatGPT's features that make the application a personal assistant as well as productivity-driven utility help to manage tasks, increase organization, and offer instant information. At the same time, it has to be well understood that despite all the "smart" solutions human factor is still a crucial need for judgment and decision-making when it comes to tasks that involve reasoning or evaluation of the information.

8.12 ChatGPT in Agriculture

ChatGPT is revolutionizing the practice of agriculture through the provision of fresh ideas and assistance in different areas of the business. Ranging from boosting crop production through precision farming and helping farmers with pest control or in-market price evaluation, ChatGPT brings precision to farming. In this way, it helps farmers make forecasts, use the provided educational material, and follow the weather conditions to act efficiently and solve problems. No matter whether it relates to optimizing daily plans of irrigation, crop sequencing, or receiving practical advice from field specialists, ChatGPT presents a multifaceted tool that assists farmers in meeting modern challenges and achieving advanced rates of development for agriculture.

8.12.1 Precision Farming

ChatGPT helps in precision farming by analyzing data from sensors and Satellite imagery to inform the farmer on the general health of crops, the condition of the soil, and the prevailing weather conditions. People can use it in farming to gain decision-making tools that enable them to maximize crop production as well as use available resources.

- **Example**: A farmer makes soil moisture measurements and asks ChatGPT for advice on the schedule of watering to save water and contribute to the plants' productivity.

8.12.2 Pest and Disease Management

Using the reports and data of pest and disease attacks, ChatGPT assists farmers in discovering threats to their plants. It gives details on symptoms, management, and possible ways of preventing the occurrence of the disease.

- **Example**: A farmer starts seeing peculiar signs in his or her produce. Pest control and management action plans as well as other related disease controls are recommended and explained by ChatGPT.

8.12.3 Crop Planning and Rotation

In particular, using historical data and the current market situation, ChatGPT is capable of suggesting the best directions for crop rotation. This makes it possible to maintain the fertility of the soil and, at the same time, make optimal profits by choosing crops that would sell in the market.

- **Example**: A farmer's activities toward the next planting season. ChatGPT provides the client with a rotation plan that will enable them to have the right crop at the right time in terms of soil health and market prices and hence efficient use of the clients' land and good profits.

8.12.4 Weather Forecasting and Advisories

ChatGPT: Weather Forecast and Alerts—Farmers can plan their activities depending on the type of weather as indicated by the system. It can give guidelines on how best to avoid or cope with severe weather conditions and how to modify the operation of cultivation.

- **Example**: He gives a weather forecast of a location that indicates that there will be heavy rain soon. This improved the farmer's access to advice on how to change the schedules of the planting and the ways of preventing damage to the crops.

8.12.5 Market Analysis and Pricing

In the business, it helps with the analysis of market trends and pricing information and advises on its application in helping the farmers sell their products at the appropriate time and place. It can give a directive possibility to determine the market demand and the change in prices.

- **Example**: A farmer consults with others regarding the most appropriate time to market his or her crops. ChatGPT determines the present prices within the market, and the timing of the sale to generate the highest possible revenue.

8.12.6 Farm Management and Automation

To recap, ChatGPT can provide recommendations on the methods of managing farms and applying technologies to automate the processes. It informs farmers on the best techniques and innovations that can be adopted to enhance production in the farm.

- **Example**: A farmer wants to implement the use of technology in irrigation. ChatGPT generates content on available technologies, their advantages, and ways of integrating the technologies.

8.12.7 Educational Resources and Training

Through the resources, tutorials, training documents, videos, tutorials, and teachings on many different methods of farming, ChatGPT contributes toward the general support of agricultural learning. This way it assists the farmers to be in touch with the recent advancement and the recommended practices to be adopted in the field.

- **Example**: A new farmer is one who wants to be taught on how to farm without harming the environment. To assist them, ChatGPT has sets of educational tools and training session that can be accessed by anyone interested.

8.12.8 Problem-Solving and Troubleshooting

When some issues arise in farming, ChatGPT responds to the issue and provides solutions based on the input information. It aids in solving problems about practices to be used in farming, equipment used in farming, and in crop management.

- **Example**: One day a farmer has issues with a piece of equipment used in farming. In response to the problem, ChatGPT offers a set of actions toward solving the problem and advises on how the issue can be handled in order to prevent future occurrence.

8.12.9 Community and Networking

Having said that, through ChatGPT, Agronomists, farmers, researchers, students, and every stakeholder in the agricultural sector can be connected in the same way they could use any other social media platform thus ensuring that they get the necessary support from experts or even their fellow farmers. It also assists in enhancing the spread of knowledge and experiences to deal with similar difficulties.

- **Example**: The farmer will consult on organic farming practices. The .AspNet provides them with the means to connect to specialists as well as other farmers who have done organic farming before, thus being a source of lessons.

8.13 ChatGPT in Travel and Tourism

In the travel and tourism segment, ChatGPT is revolutionizing how travelers begin their trips and even engage with them. From offering tailored tips for ventures, creating well-structured and scheduled travel programs, and suggesting booking options, ChatGPT works as a reliable travel companion. Many of the questions that may come up concerning travel information, including knowledge about certain destinations, cultures, and travel tips, are provided within this site. This adds value to travel by making it easier for people to plan their trips based on what they would like to do and where they would like to go thus making the whole travel a more pleasant trip for the travelers.

8.13.1 Destination Recommendations

ChatGPT can recommend places for traveling according to the client's hobby, estimated budget, and previous experience. For instance, if a user wants to travel to a beach area, ChatGPT can suggest general tourist traps as well as off-the-beaten-track places and offer information on various tourist attractions.

- **Example**: An example of the interaction is a user seeking a family-friendly vacation might inquire from ChatGPT. This way, ChatGPT could recommend attractions that are equipped with the facilities families with children might seek such as amusement parks or hotels with entertainment for children.

8.13.2 Travel Itineraries

Travelers can use ChatGPT to outplan their activities, which includes activities to be done, accommodation, and transport means to be used. It can provide users with daily itineraries based on users' preferences that include places of interest, restaurants, and attractions.

- **Example**: A tourist going to Paris could seek guidance from ChatGPT and get a detailed plan for each day of his or her stay in Paris, giving directions to famous tourist attractions such as the Eiffel Tower, delicious and catering eating joints as well as fun activities in the evening.

8.13.3 Booking Assistance

To my knowledge, ChatGPT, as a standalone software, does not make direct bookings for flights, hotels, or other travel-associated services, but it offers users useful information on how to do it. It can also recommend self-sourcing sites and applications for booking.

- **Example**: A user counseling to book a flight might turn to ChatGPT and ask for suggestions on the websites to use to book flights, or the websites to use to get the best offers, respectively.

8.13.4 Answering Travel-Related Queries

In terms of travel, ChatGPT can handle most of the inquiries that one might have depending on their destination such as visa and entrée requirements, customs, and travel warnings. It is an informative tool to assist travelers in getting through their trips with fewer challenges.

- **Example**: A person planning a trip to Japan may wish to know about the etiquette or language in the country to observe during the trip. There are features of the local people's behavior, for example, how to behave in the temple, how to use public transport in Japan, etc.

8.14 Related Work

Here are some of the related works which have contributed in the field:

- **Osama** [4]: *"ChatGPT's influence on customer experience in digital marketing: Proposing 'moderating roles' to examine."*
  The trends of digitalization make digital marketing essential to organizations. Chatbots, particularly ChatGPT, are popular for improving customer relations and making interactions more individualistic. ChatGPT, an AI-based language model by OpenAI, has been widely adopted in digital marketing due to its human-like responses. This study examines ChatGPT's role in customer experience within digital marketing, focusing on factors like business type, technology adoption, age, and education as moderators. These factors were found to significantly influence customer satisfaction, though gender had no effect.
- **Zhang** [5]: *"Co-creating with ChatGPT for tourism marketing materials."*
  With advancements in AI and machine learning, AI-generated content has revolutionized traditional marketing techniques. This work explores the influence of ChatGPT in tourism marketing, especially in generating content comparable to human marketing communication. Two online experiments conducted in Norway show that content created with ChatGPT can effectively increase tourism interest, improving marketing campaign outcomes. However, concerns arise regarding the credibility of AI-generated marketing information and its impact on the tourism marketing workforce.

- **Bencheikh and Höglund** [6]: *"Exploring the efficacy of ChatGPT in generating requirements: An experimental study."*
  This research compares AI tools, including ChatGPT, for software requirements' generation. The study demonstrates that ChatGPT can mimic human competence in generating essential project requirements, though the quality of human-generated requirements, especially from experienced participants, remains superior. The study also highlights the time-saving benefits of ChatGPT. Moreover, it distinguishes between the paid and free versions of ChatGPT, with the paid version offering more consistent and higher quality outputs.
- **Garg et al.** [7]: *"Exploring the role of ChatGPT in patient care (diagnosis and treatment) and medical research: A systematic review."*
  This review expands on the research exploring ChatGPT's potential in clinical and laboratory diagnosis, research planning, and patient care. It identifies several ethical and legal concerns, such as copyright infringement, transparency in AI content, and the implications of incorporating ChatGPT into healthcare. The review also discusses knowledge gaps, including the accuracy of ChatGPT-generated health information, legal and moral implications, AI interpretability, and data privacy issues.
- **Jenna et al.** [8]: *"Young Children's Creative Storytelling with ChatGPT vs. Parent: Approaches to interactive styles."*
  This study explores ChatGPT's role in fostering creativity in storytelling among children aged 5–6, contrasting it with parent–child interaction. In the study, eight child–parent pairs participated, and it was found that children had shorter and less frequent interactions with parents compared to ChatGPT. The conversational agent gave more positive feedback and asked different types of questions, leading to more children preferring ChatGPT's interactions. This work provides initial findings on how ChatGPT may help enrich creativity in family storytelling.

8.15 Summary

ChatGPT has shown its possibilities and practicality in various fields and it can be seen as a product with the ability to revolutionize occupations and improve many processes. Therefore, in business communication and customer relations, ChatGPT increases interaction and support effectiveness; in writing, publishing, and marketing, it optimizes the writing and marketing of content. The tool also helps in software development and technical support since it involves code and problem-solving. When it comes to data input and processing, ChatGPT helps organize, sort, and interpret huge amounts of data for definite conclusions and constant improvement. In healthcare, its usefulness is in providing accurate medical information while in the world of business, it is helpful in the study of the consumer and the market. Besides, ChatGPT is involved in creative writing and storytelling by offering ideas and help to the writers. Its major usage includes; in education, ChatGPT provides customized coaching and assistance, and in legal and compliance, it assists in investigation and compliance. The application scenarios that can be seen further explain the flexibility of ChatGPT in terms of various requirements and enhanced performance in different sectors. Moving ahead there are more possibilities and versatile opportunities that can be achieved and found by using ChatGPT. Some of the issues that may emerge in the future include ethical questions, handling large volumes of data, and data privacy, among others. It would therefore be relevant to look at how ChatGPT can be optimized once the issues of technology have been effectively handled. In total, ChatGPT is an efficient program that has various uses and can provide significant benefits to a vast number of fields and develop further progress in the following areas.

8.16 Multiple-choice Questions

In this section, you'll find a series of multiple-choice questions designed to test your understanding of key concepts in generative AI. Choose the correct answer for each question.

1. ChatGPT can perform surgeries. (True/False)
2. What is one way ChatGPT enhances business efficiency?
3. ChatGPT helps in _____ by providing personalized tutoring and support.
4. Match the following use cases with their descriptions:

Matching Exercise	
Use case	Description
ChatGPT in customer service	(A) Automates responses and handles queries
ChatGPT in content creation	(B) Generates text and marketing materials
ChatGPT in healthcare	(C) Assists in diagnosis and patient support

5. Explain how ChatGPT contributes to creative writing and storytelling.
6. Given a scenario where a company needs to automate customer support, how could ChatGPT be implemented to improve efficiency?
7. Compare the use of ChatGPT in business and healthcare. How are the applications similar or different?
8. List three pros and three cons of using ChatGPT in software development.
9. Review a case study where ChatGPT was used in a marketing campaign. What were the outcomes and lessons learned?
10. Discuss the ethical considerations and data privacy concerns when using ChatGPT in various domains.
11. If a company faces issues with customer engagement, how can ChatGPT be used to address these issues effectively?

12. Label a diagram showing the different applications of ChatGPT in various domains.
13. Select all the areas where ChatGPT can be applied from the following list:

    - (A) Business efficiency
    - (B) Hardware manufacturing
    - (C) Content creation
    - (D) Legal advice

14. Rank the following ChatGPT applications in order of impact on customer service:

    - (A) Automated responses
    - (B) Data analysis
    - (C) Personalization
    - (D) Feedback collection

8.17 Answers

Below are the answers to the multiple-choice questions from the previous section:

1. False
2. By automating customer support and streamlining communication processes
3. Education
4. Matching:

    - **Content Creation**: (c) Assists in creative processes and marketing strategies
    - **Healthcare**: (a) Provides reliable medical information
    - **Market Research**: (b) Analyzes consumer behavior and trends
    - **Data Entry**: (d) Manages and interprets large datasets

5. ChatGPT contributes to creative writing and storytelling by generating ideas, suggesting plot developments, enhancing dialog, and providing feedback on character development, thus facilitating the creative writing process.
6. ChatGPT can be implemented by creating a chatbot that handles common customer inquiries, provides instant responses, and escalates complex issues to human agents for resolution.
7. ChatGPT enhances efficiency in both business and healthcare by providing relevant information and automating processes. Business applications focus on customer engagement

8.17 Answers

and operational efficiency, while healthcare applications prioritize accuracy and quality of medical information.

8. **Pros**:

   - Assists in generating code snippets quickly.
   - Helps with debugging and troubleshooting issues.
   - Provides immediate responses to programming queries.

   Cons:

   - May produce incorrect or inefficient code.
   - Limited understanding of complex project requirements.
   - Over-reliance on AI could reduce developers' problem-solving skills.

9. Outcomes may include increased engagement rates and enhanced content generation efficiency. Lessons learned could involve the necessity of fine-tuning the AI for specific campaign needs and the importance of human oversight to address nuanced situations and ensure content accuracy.
10. Ethical considerations include ensuring that ChatGPT does not reinforce biases and is used responsibly. Data privacy concerns involve safeguarding user data and complying with regulations such as GDPR to protect user information and ensure transparency.
11. ChatGPT can enhance customer engagement by providing personalized interactions, offering instant responses to inquiries, analyzing engagement data to refine strategies, and delivering tailored content based on user preferences.
12. (Labels will depend on the specific diagram, but examples could include: Business Efficiency, Healthcare, Content Creation, Data Analysis.)
13. **Correct Areas**:

    - (A) Business efficiency
    - (C) Content creation
    - (D) Legal advice

14. **Ranking**:

    - Automated responses
    - Personalization
    - Data analysis
    - Feedback collection

References

1. Abid Haleem, Mohd Javaid, and Ravi Pratap Singh. Exploring the competence of chatgpt for customer and patient service management. *Intelligent Pharmacy*, 2024.
2. Alec Pappas, Elena Fumagalli, Maria Rouziou, and Willy Bolander. More than machines: The role of the future retail salesperson in enhancing the customer experience. *Journal of Retailing*, 99(4):518–531, 2023.
3. Ethan Dickey, Andres Bejarano, and Chirayu Garg. Ai-lab: A framework for introducing generative artificial intelligence tools in computer programming courses. *SN Computer Science*, 5(6):720, 2024.
4. Osama A. Abdelkader. Chatgpt's influence on customer experience in digital marketing: Investigating the moderating roles. *Heliyon*, 9(8):e18770, 2023.
5. Author Zhang. Co-creating with chatgpt for tourism marketing materials. *Annals of Tourism Research*, 96:100124, 2024.
6. L. Bencheikh and N. Höglund. Exploring the efficacy of chatgpt in generating requirements: An experimental study, 2023.
7. RK Garg, VL Urs, AA Agarwal, SK Chaudhary, V Paliwal, and SK Kar. Exploring the role of chatgpt in patient care (diagnosis and treatment) and medical research: A systematic review. *Health Promotion Perspectives*, 13(3):183–191, 2023.
8. Jenna H. Chin, Seungwook Lee, Mohsena Ashraf, Matt Zago, Yun Xie, Elizabeth A. Wolfgram, Tom Yeh, and Pilyoung Kim. Young children's creative storytelling with chatgpt vs. parent: Comparing interactive styles. In *Extended Abstracts of the 2024 CHI Conference on Human Factors in Computing Systems (CHI EA '24)*, pages 1–7, New York, NY, USA, 2024. Association for Computing Machinery.

The manufacturer's authorised representative in the EU is Springer Nature Customer Service Centre GmbH, Europaplatz 3, 69115 Heidelberg, Germany. If you have any concerns regarding our products, please contact ProductSafety@springernature.com

Printed and bound by CPI Group (UK) Ltd, Croydon, CR0 4YY

10/03/2026

02068980-0014